Industrial Technical Illustration

JON M. DUFF The Ohio State University

Industrial Technical Illustration

VNR VAN NOSTRAND REINHOLD COMPANY
NEW YORK CINCINNATI TORONTO LONDON MELBOURNE

Published by Van Nostrand Reinhold Company Inc.
135 West 50th Street, New York, N.Y. 10020

Van Nostrand Reinhold Publishing
1410 Birchmount Road
Scarborough, Ontario M1P 2E7, Canada

Van Nostrand Reinhold Australia Pty. Ltd.
17 Queen Street
Mitcham, Victoria 3132, Australia

Van Nostrand Reinhold Company Limited
Molly Millars Lane
Wokingham, Berkshire, England

15 14 13 12 11 10 9 8 7 6 5 4 3 2 1

Library of Congress Cataloging in Publication Data

Duff, Jon M., 1948–
   Industrial technical illustration.

   Bibliography: p.
   Includes index.
   1. Technical illustration. I. Title.
T11.8.D83 1982b   604.2'4   82-8315
ISBN 0-442-21957-1   AACR2

Because technical illustration comprises several kinds of technical drawing, this book is titled *Industrial Technical Illustration* to differentiate it from books in the related areas of medical, scientific, and architectural illustration. This book presents a highly organized approach to learning technical pictorial drawing. It contains specific instruction in orthographic or multiview drawing and in a direct method of parallel pictorial drawing called axonometric drawing. In my view, axonometric pictorial drawing is the most important skill that any industrial technical illustrator can master.

In the past, industrial technical illustrations were done almost solely from engineering drawings or from blueprints. This practice is still common. However, as the result of microfilm image requirements, skyrocketing labor costs, and the development of computer-aided pictorial drawing, much illustration is now done with the aid of any of several mechanical or electronic devices. Illustrations are currently done from rapidly drawn computer underlays called blockouts, from photographs, or from existing drawings in a paste-up. Yet the foundation of industrial technical illustration remains the ability to read and understand engineering drawings. Thus, the student is urged to acquire as great a skill as possible in reading engineering drawings and in translating these drawings into descriptive pictorial images.

Technical illustration relys on sound knowledge of technical drawing techniques. Therefore, *Industrial Technical Illustration* assumes a working knowledge of orthographic or multiview drawing, sectioning, auxiliary views, dimensioning, and the reading of engineering drawings. Thus, the pictorial drawing discussed in this text is best learned *after* taking one or more courses in technical drawing.

Not all technical illustrators do the same job in the same way. Not only are there personal differences from one illustrator to another—differences in training, experience, skill, and motivation—there are differences in employers' requirements and needs. Consequently, illustrators need to be broadly trained, able to adjust to varied work requirements and flexible enough to adjust their illustration practices to conform to the needs of their employers.

Practices differ not only from one company to another but from one industry group (heavy machinery, for example) to another (such as medical equipment). Many industries have time-honored practices that the illustrator must rapidly learn. Others may not have used illustration in the past and may therefore be more open to different illustration techniques.

Methods of producing illustration vary. Even though most is produced using conventional manual drafting equipment, more and more industrial technical illustration uses mechanical and computer-assisted graphics devices to speed up the drawing process. Computers don't produce finished technical illustration—at least not yet. The illustrator still needs to know how to construct the illustration and then how to finish the drawing and make it come alive.

Companies that use technical illustration range from backward to up-to-date to thoroughly modern in terms of how they make, store, retrieve, and transmit their drawings. The illustrator must be prepared to work in any of these environments.

An accurate pictorial drawing is only the beginning of a good illustration. The drawing must use effective presentation techniques, such as line and tone, to describe each part. In this book, instruction in rendering technique is accompanied by professional hints for both students and practicing illustrators.

Continuous tone illustration, often in the form of airbrush art, is covered only briefly in this book, for several reasons. For one thing, it has nearly priced itself out of the market. Also, current reproduction requirements call for reduction and blow-back several generations removed from the original illustration.

The study problems included in this book are current and are taken directly from industry. They are organized in the manner in which an illustrator would organize them on the job. The study problems include engineering sketches, engineering drawings, computer block-outs, and photographs. A student who completes these problems will have experienced the full range of industrial technical illustration. Additional practice should help the student master entry-level job skills.

The Appendix contains, among other helpful items, a library of the hardware that is often encountered in making illustrations. Accurately constructing these hardware items helps the drawing come alive and often means the difference between a successful illustration and an unsuccessful one.

The hardware appendix also shows the pictorial construction technique used to draw each item in axonometric and its standard sizes. Practicing illustrators should find this section helpful in accurately and rapidly laying out and finishing their own industrial technical illustration.

Because students are often placed in situations where they must guess at a particular construction or avoid drawing a difficult part because they don't have the right aid or guide, *Industrial Technical Illustration* includes axonometric scales, protractors, ellipse selector charts, grids, and guides. These are the same items that are available to the illustrator on the job. The best way to use them is to photocopy the pages and then make transparencies of the copies. Each student can then have his or her own personal guides to use in solving the problems.

There are opportunities in every major industrial center for individuals who can efficiently produce technical illustrations. This book can be your first step in mastering the required skills, or it can serve as a reference book to make your illustration work a little easier.

Jon M. Duff

Credits and Acknowledgments

The following groups and individuals contributed their expertise in the form of examples included in *Industrial Technical Illustration*. Their help was greatly appreciated and contributed to the text's diversity as well as to its effectiveness in developing the reader's technical illustration skills.

Jeffrey Mining Machinery Division, Dresser Industries, Inc.
The Nelson Studio
Wes Lerdon & Associates Industrial Design

Bob Abbott, Lennox Industries
Saginaw Steering Gear Division, General Motors Corp.
Tektronics
James Shough & Associates Visuals
Ron Gardner
Kenneth Balliet
Pennsylvania Engineering Corporation
Industrial Graphics International
Society for Technical Communications

Mechanics Illustrated, The Hearst Corporation
Frank Rowe
Graphic Standard
Axograph Ltd.
Sargent and Lundy Engineers

Larry Goss of Indiana State University, Evansville, also added his teaching and illustration experience to a review of the text. His suggestions contributed to *Industrial Technical Illustration* being a text for teachers, students, and illustrators alike. His help was greatly appreciated.

I would also like to thank the following people who reviewed the text:
Harry E. Clair, Texas State Technical Institute
Laura Ann Gilchrist, Los Angeles Trade Technical School
James O'Neal, Purdue University

Contents

Industrial Technical Illustration

CHAPTER ONE The Who, What, and Why

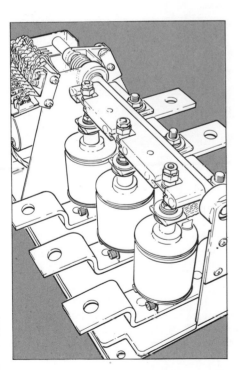

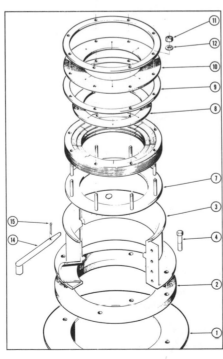

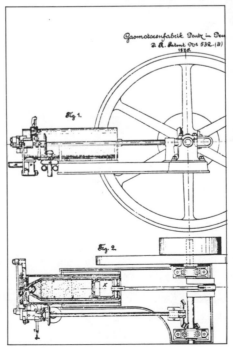

Industrial technical illustrations describe spatial relationships in a manner far superior to that of traditional engineering drawings. Because the drawing is a picture, the difficult job of "reading" the flat views is made easier. More people of varying graphic ability can understand an industrial technical illustration than can understand the engineering detail drawing (Figure 1-3). Although this type of pictorial drawing does several things quite well, it *can* be inappropriate at times. As a means of instruction for assembly or servicing, it can't be beat. Yet, however effective it is in presenting overall spatial relationships, it does so at the expense of detail. In order for the part or product to actually be made, scaled detail drawings must be used; the detail drawings are understood by trained machinists and other technical workers. The less technical training the drawing's user has, the more illustrative or pictorial the drawing must be.

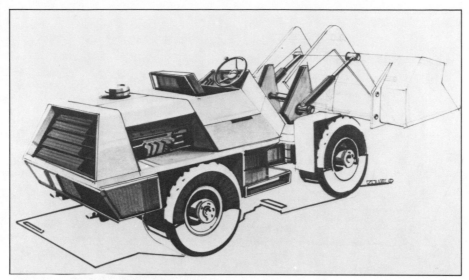

Figure 1-1. Technical illustration often describes a product as a picture, giving a more accurate representation than an engineering drawing. This illustration, done in marker and pencil, concentrates on the machine's body panels while minimizing the wheels and loading bucket. (Courtesy of Frank Rowe, Jr.)

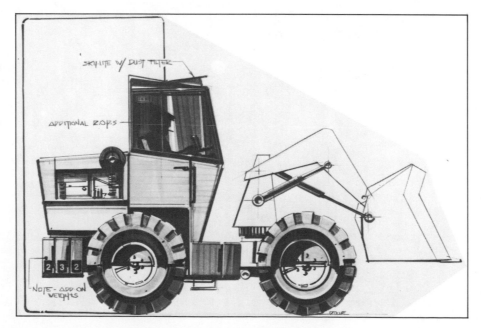

There are two main pictorial techniques available to the technical illustrator: (1) parallel pictorial drawing, often called *axonometric drawing* (Figure 1-4), and (2) converging pictorial drawing, called *perspective drawing* (Figure 1-5). Each of these is covered in detail in the following chapters. But let me offer a word of caution. Although perspective drawing produces a picture most nearly like that seen by the eye, its use in industrial technical illustration is limited. Because distances progressively become shorter as lines recede to the horizon, it is often impossible to accurately measure distances on the drawing without specific perspective scales. If the expense of producing illustrations is to be justified, technical workers need to be able to

Figure 1-2. Standard orthographic views (top, front, side) can be illustrative when shading and linework techniques are used to render the product's form. As in Figure 1-1, parts of this machine are left as simple lines. Descriptive text is added as necessary. The rectangular form behind the machine is called a *panel*; it helps pull the object forward, separating the machine from the background. (Courtesy of Rank Rowe, Jr.)

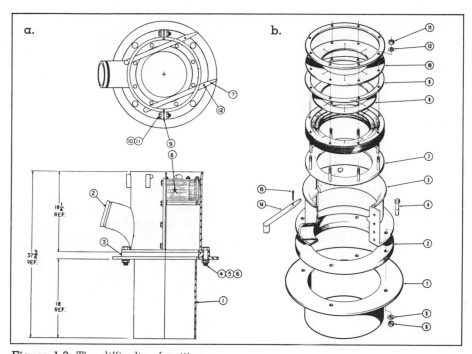

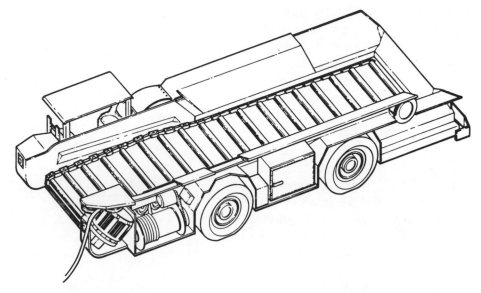

Figure 1-3. The difficulty of getting a clear picture from orthographic engineering drawings is apparent from this example. Compare the engineering drawing (a) with the pictorial assembly drawing (b). In (a), multiple views are needed to assemble the three-dimensional quality of each part. Even worse, the way that the parts fit together is difficult to understand. In (b), not only can the shape of each part be seen, but the way that each part fits into the next is obvious and the sequence of assembly is easier to understand. A person with limited technical skill would have difficulty understanding the engineering drawing, but almost anyone could understand the pictorial approach.

Figure 1-4. A parallel pictorial drawing presents the subject realistically, as long as parallel lines—lines that would normally converge as they recede into the distance—are not allowed to visually "spread." The object is usually oriented so that the part of greatest interest is on the long axis whereas the less important side is on the short axis.

Figure 1-5. Perspective drawings are less "stiff" than parallel pictorial drawings, as this example shows. The parts of the object that are farther away from the viewer are at a different scale than those closer to the viewer, even if they are aligned along the same axis. When deciding whether or not to make the illustration a perspective, ask this question: will the drawing be used to coordinate engineering activities or will it need to give a more realistic picture? Perspective drawings may be useful for marketing; for engineering, probably not.

Although perspective represents a small portion of industrial technical illustration, several companies continue to use specialized perspective drawing. In such cases, views are standard, with common points identified so that perspective scaling can be used.

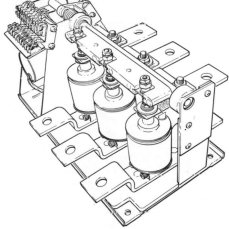

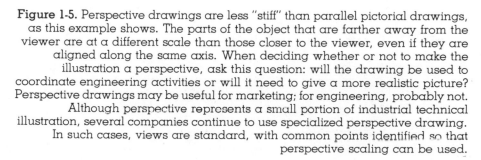

use them without difficulty. An engineer should be able to measure, or at least compare, distances between hole centers, for example, without using elaborate measuring devices. The drawing technique used to avoid this difficulty is the parallel or axonometric pictorial drawing. It gives an accurate, easily scaled, visually representative picture of the object (Figure 1-6). Every illustrator should be able to draw it in its three forms—isometric, dimetric, and trimetric.

Is perspective drawing a waste of time? Certainly not. There are times when only a perspective image will show an object just the way you want it. In fact, perspective drawing helps the illustrator understand axonometric drawing (Figure 1-7).

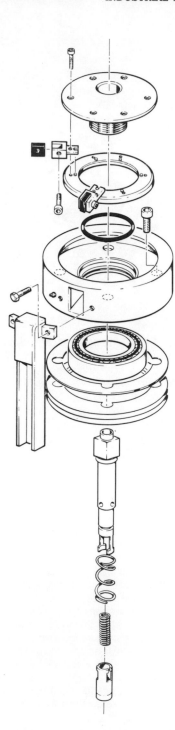

Figure 1-6. An axonometric illustration should build a truly three-dimensional picture of an object. This 3-D effect is achieved by the selective use of different line weights, by the placement of lines, and by the placement of parts in relation to each other.

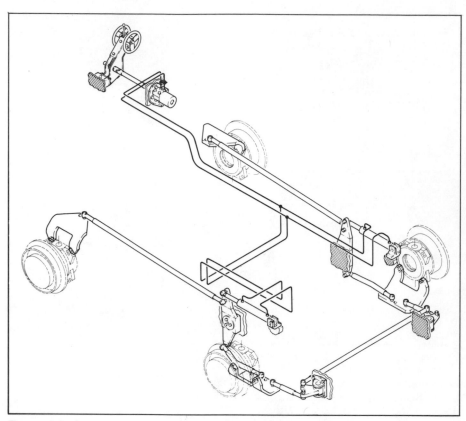

Figure 1-7. An experienced illustrator will be able to choose among many techniques in order to draw and then render an object. These techniques include sectional practices, phantom views, and the simplification of certain components. In this example, knowledge of perspective aided in keeping the parallel axonometric lines from "spreading" as they moved further away.

A Short History of Technical Illustration

The birth of technical illustration had to wait for theories of drawing to emerge and become well developed. The early profile drawing of the Egyptians and the plan drawings of the Greeks could not combine in one view the height, width, and depth of an object.

Around A.D. 1500 several factors came together to bring about the development of technical illustration. At this time, pictorial drawing took on a truly three-dimensional quality. What happened inside the eye was accurately translated onto the surface of the paper. This period also marked the beginning of man's interest in the technical aspects of his world, his body, and his machines. The most famous of the group of thinkers and illustrators of the time was Leonardo daVinci. He is a model for illustrators even to this day because he was an accomplished artist, sculptor, inventor, thinker, and natural philosopher (Figure 1-8). His artistic ability and scientific questioning provided the means and the stimulus for descriptive technical illustration.

The Industrial Revolution further refined the field of technical illustration. Standards were set, conventions were accepted, and techniques were developed. Yet the drawings remained without real three-dimensional depth, appearing more like orthographic views (Figure 1-9).

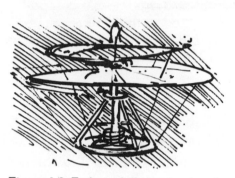

Figure 1-8. Technical illustration has its historical foundations in the artists of the 15th and 16th centuries. The most visible was the artist, philosopher, inventor, and natural scientist Leonardo da Vinci. He used drawing as a way of describing technical, medical, and engineering principles. He also used drawing as a thinking tool. This redrawing illustrates one of his investigations—the operation of a helicopter.

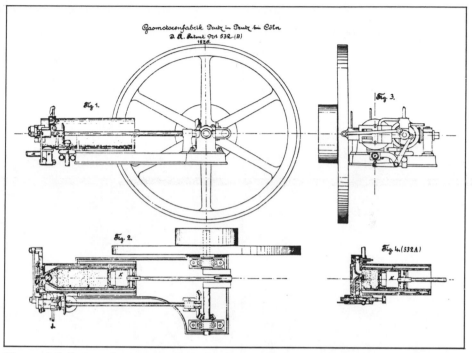

Figure 1-9. Patent drawings reflect the illustrative engineering drawing practices of the 19th century, even to this day. The technical accuracy of the design is maintained—for obvious reasons—yet illustration techniques give a more realistic view of the object than would be given with standard engineering drawing practices. In this reproduction of N. A. Otto's first patent application, several current illustration techniques have been used, including line shading on the flywheel, line weight changes to show solidity, and selective use of hidden lines.

Most engineering drawings exhibited this "engraving" look in the period 1840–1910. It reflected the accepted drawing practices, drawing instruction, and reproduction technology of the time.

Note that the two views to the left (Fig. 1 and Fig. 2) appear to be out of projection. Also, Fig. 3 looks like it should be on the left side. This reflects the practice in some countries (especially in Europe) of arranging the orthographic views differently from the method generally used in the United States.

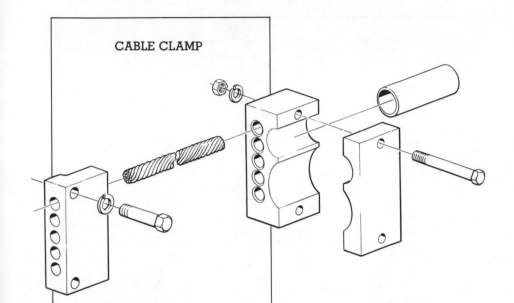

CABLE CLAMP

TO ECONOMIZE IN A DRAWING, NOT ALL PARTS NEED TO BE SHOWN. ONLY ONE OF THE PARTS IS DRAWN, AND THE EXACT NUMBER REQUIRED IS NOTED IN A BILL OF MATERIALS. HERE, THE BLOCK ON THE LEFT IS CONNECTED TO THE SPLIT BLOCK ON THE RIGHT BY FIVE LENGTHS OF STEEL CABLE. ONLY ONE OF THESE CABLES IS SHOWN AND IS BROKEN SO THAT THE ASSEMBLY CAN BE SHOWN IN PROXIMITY.

Jumping to more recent history, most would agree that industrial technical illustration as it is currently practiced developed in the period 1940–1955 and became increasingly refined with the boom in commercial air travel and with the space program. Illustrators were trained by the major aircraft and automobile companies. Today these illustrators have been absorbed into almost all manufacturing, construction, and service industries.

Who Are the Illustrators?

Industrial technical illustrators are either artists with sound technical background or technically trained people who enjoy visual as opposed to verbal or mathematical description. They produce drawings that aid in product development, marketing, assembly, and service. They must understand the language of engineers, printers, photographers, and machinists.

Artists without a basic knowledge of materials, processes, manufacturing, and the principles of industry and engineering would find it difficult to make drawings that capture the technical relationships of an object. Likewise, nonartistic technicians, technologists, and engineers may be able to "read" engineering drawings but would be unable to translate them into pictures.

Illustrators may be employed as part of an engineering staff within a company, as members of a separate technical-services (technical publications) department, as artists in a technically oriented art agency, or as private freelance or subcontract employees. As those trained in the 1940–1955 period begin to retire, a whole new crew of illustrators will surface. It is to them—you—that this book is directed.

What Is Illustration?

The industrial form of technical illustration is produced in any of four ways: from engineering sketches, from engineering drawings, from photographs, and from computer-generated underlays.

FROM ENGINEERING SKETCHES

The engineer, who isn't an illustrator, often produces sketches of designs or situations. They are usually crude—surely not the kind of drawing one would include in a book or manual (Figure 1-10). These sketches are produced when a project is in its very early stages, when management wants to see what something will look like but is still unwilling or unable to become more deeply involved in the project. The illustrator must use the engineer's rough notes and, with his own technical knowledge, assemble a picture of what the engineer has in mind. To say the least, it takes a great deal of experience to do this. An illustrator just starting out would not have the in-depth knowledge to put in the detail that the engineer expects.

FROM ENGINEERING DRAWINGS

To be accurate, an illustration must be based on detailed and accurate information—the more complete and accurate the information, the better. Engineering documents (the bill of material, specifications, assembly drawings, details, and sections) provide the kind of in-

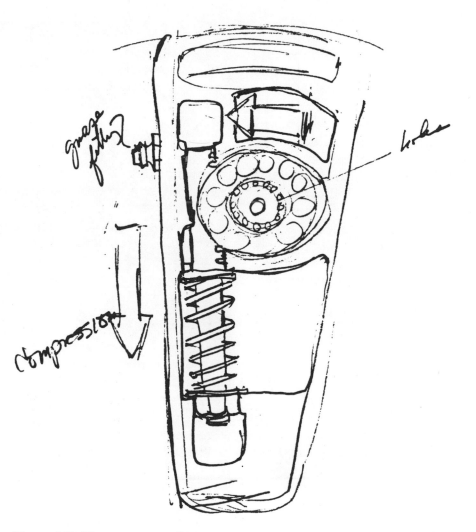

Figure 1-10. The engineer will often produce a quick sketch, establishing the functions of an object and defining the shape so that an illustrator can make the drawing more descriptive. Sometimes the engineer will add that the part "looks like Model 5739-B, but longer and beefier," or something to that effect. The illustrator must be ready to research the problem, gather the required information, and add his work to that of the engineer. Since the illustrator isn't working from detail drawings, he should get his pencil layout approved before going on. This precaution can save having to redo a finished drawing at a later time.

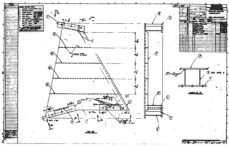

Figure 1-11. Engineering drawings contain a wealth of information on size, shape, fit, material, strength, and assembly. However, this information is presented according to complex "conventions"—standardized ways of showing these properties of the object. To make things more difficult, the standards differ, depending on whom you are working for and what country they are used in. Standards also vary in accuracy.

People "read" engineering drawings just like they read a book, only instead of reading words they must read pictures. In fact, a set of engineering drawings is much like a puzzle, with information on one drawing leading the reader to other drawings and back again. Well-made engineering drawings are fun to use. Poorly made drawings can make an illustration job very difficult.

formation needed to produce effective and accurate illustration.

Engineering drawings (Figure 1-11) exist today much as they have for three thousand years. Yet the key for rapidly understanding their parallel orthogonal positioning of views has yet to be found. Some people learn this "reading"

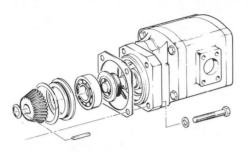

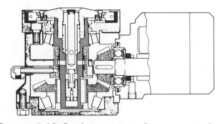

Figure 1-12. In the sectional view, much more information is provided than would be needed to make a pictorial illustration. How much detail will show up on the final pictorial illustration? An experienced illustrator can determine when to include fine detail and when to leave it out.

easily, whereas others learn it only with difficulty. Many who wish to learn never find the key. Illustrators *must* be able to read engineering drawings.

Although it is useful to have all the details and information on which to base an illustration, there is so much information in an engineering drawing that it is a big job for the illustrator to sift through it, sorting out what he or she needs to solve each drawing problem. Much of the ability to quickly find what you need comes with experience.

FROM PHOTOGRAPHS

There is a place for photography in technical communications—a big place (Figure 1-13). Yet even the most skillful photographers have difficulty controlling the many factors that make a technically accurate photograph difficult to produce and reproduce. Without extensive retouching, most photographs are unacceptable as illustration. But photographs can relieve the illustrator from the task of laying out a drawing, especially in cases where no engineering drawings exist for the subject. Using photos also makes sense when the part or installation is so complicated that a constructed drawing

Figure 1-13. Technical photography goes hand-in-hand with technical illustration. They should complement each other rather than compete. A rule I like to adhere to is: Don't make illustration look like photography . . . and don't use photography as a substitute for illustration. What this means is that you should save photography for the things it does well, such as establishing a visual record of temporary conditions, making microphotographs, and providing realistic depiction. Avoid using photographs because you can't draw, or drawings because you can't photograph.

would be too time-consuming. If the photographer has used what is called a "normal lens" to take the picture, the drawing will end up with a relatively undistorted perspective. In these cases, photographs serve as underlays. The illustrator traces over the photograph on a transparent medium. Such tracings made over photographs are usually known as *photodrawings* (Figure 1-14).

Any see-through material will work as an overlay. Some work better with pencil whereas

(a)

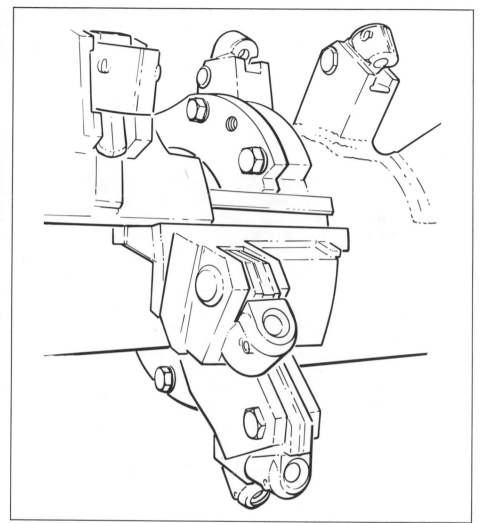

(b)

Figure 1-14. Without manipulation of the negative or print (as in photo retouching), the camera simply records what is in front of its lens. As in this example, that often includes things you wouldn't necessarily want in the photograph—metal cuttings, chalk marks, oil drips. A photodrawing can be produced at a fraction of the cost of retouching the photograph, with another advantage: the illustration is line copy and can be reproduced without a halftone screen.

others take ink. It sounds easy, but the trick is to know what and how much of the photograph to leave out. The illustrator has to be selective in what to show, what to accentuate, and what to minimize.

FROM COMPUTER-GENERATED UNDERLAYS

Computer-generated perspective and axonometric drawing has been around for some time, starting with the hard-wired analog machines, in which engineering drawings were traced over with a pointer, and progressing to three-dimensional digital computer graphics systems. Up until recently, even the most expensive programs could not produce reproduction-quality illustration. What they *could* do was produce accurately scaled underlays or blockouts of the general pictorial construction. It was still up to the illustrator to trace over the computer drawing.

Current computer illustration provides pictures that can be redrawn rapidly enough that the operator can revolve the object into its most favorable orientation. Computer programs delete hidden lines, vary

line weights according to a predetermined scheme, and keep standard or often-needed parts such as bolts, nuts, springs, bearings, and pins in computer memory, much like a template. Thus, the operator doesn't have to redraw these parts every time they appear on the drawing.

Computer illustration does not mean an end to individual illustration. In fact, the more that a person knows about drawing, the more he or she can use devices such as computers as labor-saving tools. It takes a substantial volume of similar-product illustration to justify the use of computer systems. When this use is justified, the illustrator must be ready, and willing, to accept the computer as a fast, smart pencil. The human computer is constantly learning, continuously changing. Nonhuman computers are not. Humans aren't paid when they go home. A computer-graphics system costs you every minute that it sits there—and double for every minute it sits idle.

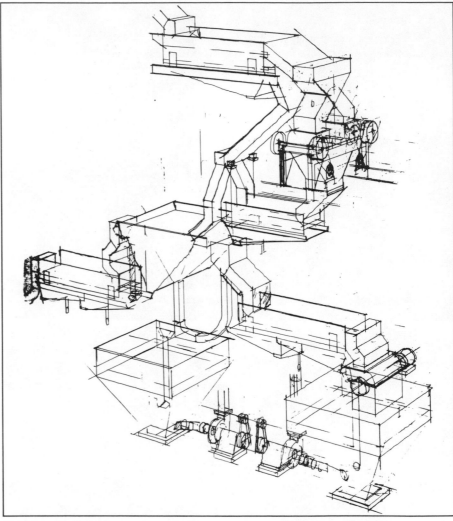

Figure 1-15. The computer underlay in this example was produced from engineering detail drawings with the aid of an Illustromat. Note that the lines and ellipses are approximate—they will be cleaned up on the final drawing. The engineering drawings were manually traced with the machine, and with instruction from the operator, the machine automatically scaled the axes and tilted the object.

The operator has to be selective in how much of the object to trace. The machine cannot decide what is on the back side of the object and therefore will not be seen. It is up to the operator to stop tracing at the right time to make the object look solid. (Courtesy of Nelson Studio, Inc.)

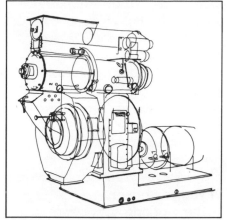

(a)

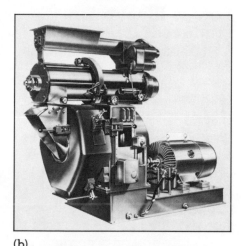

(b)

Figure 1-16. There must, of course, be much finishing of a computer blockout in order to produce a final illustration. In this example, a rough blockout has been used to make a finished airbrush illustration. Lines have to be straightened, ellipses smoothed with standard guides, and detail added from engineering drawings or photographs. The finished illustration is a combination of brush, airbrush, and pen. (Courtesy of Nelson Studio, Inc.)

Where Is Illustration Used?

Illustration is used at various points throughout the course of developing an idea or product. Many would say that illustration is used to communicate. Well, yes, of course it is used to communicate, but so are many other human activities. This statement is like saying that a person is tall without telling you *how* tall, tall compared to whom or what, and according to what

measuring device. Illustration communicates but in distinctly identifiable ways:

1. During design of a product, illustration is used in a freehand manner to check the arrangement of components (Figure 1-17).
2. Illustration is used for making documents or drawings so that the product can be built or assembled (Figure 1-18).
3. Illustration is used for persuading others to purchase the product and for telling them how to use, service, or dispose of it (Figure 1-19).

Figure 1-17. Industrial technical illustration can help a designer explore the possible visual solutions to a problem. A designer will go back and forth between orthographic and pictorial drawings, trying out possibilities and attempting to sense the appropriateness of different solutions. The illustrations that are produced in this process are less formal, more open to change.

The designer not only communicates with himself but also with peers, supervisors, and clients. The designer's illustrations—if they are going to work—must accurately reflect the ideas that are inside his head. Otherwise, the illustrations are unsuccessful—they don't work.

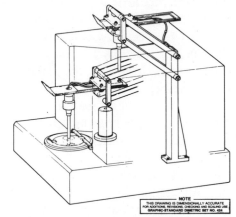

Figure 1-18. Illustration can be used in place of engineering drawing if, for some reason, the engineering drawings cannot be produced. In such cases great care must be exercised to maintain the technical and dimensional accuracy of the drawing. A technically accurate illustration may be pinned up on a wall to serve as the source of information for the building of a prototype.

One company, Graphic Standard of Troy, Michigan, provides labels so that revisions or checking will be based on the built-in accuracy of drawings produced with the aid of their axonometric scales and guides.

Figure 1-19. Illustration aids in the servicing of complex components. In this airbrush example, certain parts have been shown in section while others have been treated with a phantom technique so that more important detail may be shown. (Courtesy of Lennox Industries, Inc.)

TRAM CASE

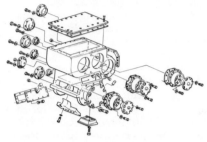

EVERY ILLUSTRATION IS
A COMPROMISE. NOT
ALL DETAIL CAN BE
SHOWN AND STILL KEEP
THE DRAWING A
MANAGEABLE SIZE.
THE ILLUSTRATOR HAS
TO ASK "WHAT AM I
TRYING TO SHOW?" PARTS
TO THE FRONT CAN EASILY
BE SEEN; PARTS TO THE
REAR CAN BECOME
CONFUSED. AND SPLITTING
THE ASSEMBLY INTO
TWO DRAWINGS CAN
TURN A 200-PAGE
PARTS BOOK INTO A
350-PAGE PARTS BOOK.

The illustrator can handle the drawing differently depending on its intended use. Most teachers, and all professional illustrators, have encountered an illustration that looked good but still didn't do what it was intended to do. This type of failure can often be traced to the fact that the illustrator produced the drawing for the wrong use. Remember, illustration is important not for how it is produced but for how it aids in designing, documenting, and persuading.

Because of this need for versatility, this book is not broken down into assembly drawing, section drawing, or detail drawing. Instead, *views* of technical subjects are produced to show the objects in a particular orientation. The orientation may be orthogonal or perspective; the orthogonal orientation may be normal, cardinal, isometric, dimetric, or trimetric; the perspective may be angular, parallel, or oblique. When you think of technical drawing in this way, the chances of your choosing an inappropriate view are reduced. Samples of these views are provided in Chapter Five (Figure 5-3).

How Is Illustration Produced?

As mentioned before, industrial technical illustration is produced from sketches, engineering drawings, photographs, and computer underlays. Regardless of the method used, a regular sequence of events is generally followed. A step-by-step progression through the sequence will assure that the illustration is done as rapidly and as accurately as possible. Undoubtedly, you will develop your own way of organizing your work. But as a start, consider this approach:

1. Keep all letters, instructions, and notes in a folder or notebook. Keep the material in chronological order so that if

there is ever a question as to who said what or as to when something was said, you can find it.

2. Pull all drawings, photographs, bills of material, and specifications for the assignment. Organize the drawings into a "working set." Keep specifications, bills, and photographs in document protectors in your notebook.

The working set of engineering drawings represents all of the information available on the assignment. It should be kept intact, in an order that makes searching for information easy for you. In general, the set should be organized as shown below, with a plain cover sheet.

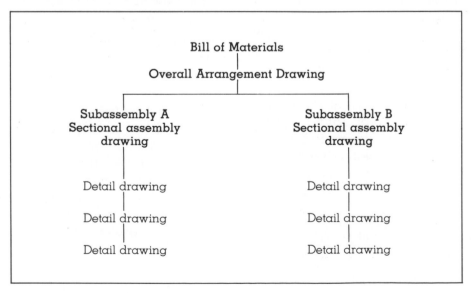

3. Familiarize yourself with the assignment. Take some time to figure out how the thing works, what the individual parts look like, and what kinds of fasteners are used. Use colored pencils to shade the sectional assemblies if you need to, and write the part names next to the call-out numbers. Do anything that will help you understand the assignment.

4. Decide on the view that will display the object in its most advantageous orientation. You may want to do some thumbnail sketches over grid paper (see Appendix D). If you are using a static computer graphics device, now is the time to decide the view. If you are using 3-D graphics with the ability to rotate the object, now is the time to enter the description of the object into the computer.

5. Determine the source or center item of the assembly, if it is made up of several parts, or choose a starting plane of an individual object. In sequence, produce the part or parts in proper projection. Don't be overly concerned with placing the parts in correct position on the page. Just make sure they are oriented correctly. You can adjust the placement of individual parts in your final tracing. If you're using a computer, output the parts for paste-up, or rotate the objects individually into position and output the entire assembly.

6. When you think you have the assignment completed, go back to the bill of materials and check off each part. Then examine each item and the corresponding detail sheet for accuracy. Make a copy of your drawing and secure appropriate approval before starting on the finished illustration.

7. When you're satisfied with the accuracy of your underdrawing, overlay tracing vellum or acetate and, using consistent presentation techniques, render the illustration.

8. If you can, make a copy of the finished tracing for your portfolio. Mount the illustration on board even if you aren't required to, and cover it with a tissue and paper flap.

9. Get ready for your next assignment by cleaning your equipment and work area. File your notebook and set of working drawings for future reference.

Just a word on speed and accuracy before we get into the how-to part of the book: the illustration of technical subject matter *must* be accurate, yet economical. With experience, an illustrator will learn when to leave out certain detail, when to take shortcuts, and when to draw a part freehand. These are all judgments that make an illustrator efficient and a valuable part of the technical team. A technical illustrator should strive to develop speed while keeping the level of accuracy required for the job.

IN ORDER TO SHOW ONE PART BEHIND ANOTHER, YOU CAN ALLOW THE FRONT PART TO DISAPPEAR, USE DASHED LINES, USE A BROKEN SECTION, OR, AS IN THIS CASE, PRESENT A PHANTOM VIEW. THE REAR PART IS SHOWN SOLID, WHEREAS THE FRONT PART FADES AWAY. HERE, BRINGING THE SHAFT FORWARD USES LESS SPACE AND ALLOWS A LARGER FINAL REPRODUCTION SCALE.

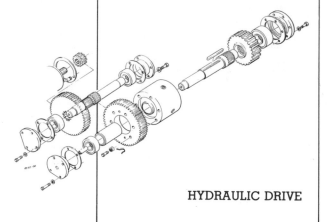

HYDRAULIC DRIVE

List of Terms

These terms were used in Chapter One. Read through them and see how many you now understand. Don't be sur- prised if several are still un- familiar to you. They will appear again in the next few chapters.

art agency
axonometric drawing
computer blockout
design, document, persuade
detail drawings
digital computer graphics
engineering department
engineering drawing
engineering sketch

freelance or subcontract artist
industrial technical illustration
perspective drawing
photodrawing
pictorial drawing
"reading" a drawing
technical services department
working set of drawings

Questions for Further Study

1. How many different jobs re- quire the worker to make tech- nical illustrations? How many more require that technical il- lustrations be used by people who didn't have a hand in making them?

2. When might a technical photograph be more effective than a technical illustration?

3. Many verbal descriptions would work better if they were translated into pictures. Find a short write-up of an operation or technique (changing batteries in a flashlight, for example) and redo it as pictures. Select a grid sheet from Appendix D and use an overlay sheet to make your drawings.

CHAPTER TWO Aids and Guides

MOV COPY	COPY ENT TO	TXT	PNT	FIG
MEAS DIST	INTOF	PRT	LDIM	MX
MOD NLIN	SET HIDN	PNT XOYO	XH	MY
DIP14	CAPD	PC.BUL	DIODE	RES
DIP16	CAP	ARS	NPN	RESV
DIP8	CAP+	VCC	PNP	CAP
RES4	DIODE	GND	IN	CAPV
RES8	TRANS	SCH.BUL	OUT	CAP+

An illustrator uses a number of aids, guides, and tools to make the job easier and faster. The longer an illustrator is on the job, the more shortcuts and standardized methods he or she discovers. Among the most useful materials are mechanical guides and templates to accurately reproduce details that repeat from illustration to illustration, grids and boards, mechanical and electronic devices, photographic and computer underlays, and a wide variety of tools.

Templates

Many manufacturers produce systems of plastic templates for use in making technical illustrations. Their catalogues contain templates for almost any application. If you need a special template and are willing to pay for it, many manufacturers will make one for you.

The most important templates for technical illustration are ellipse templates—both large and small. Although ellipses can be constructed or plotted (see Chapter Five), it would be too time consuming to do so for every ellipse. Thus, a full set of ellipse guides is indispensible.

In using ellipse templates, it is important to keep several points in mind. The ellipse is a non-normal view of a circle. As such, it has a *major axis*, which will be equal to the diameter of the circle for horizontal faces, and a *minor axis*, which will become smaller and smaller as the circle is made flatter (see Appendix C, Table C-7). The size relationship of the major to minor axis determines the exposure of the ellipse (Figure 2-1). An ellipse can easily get twisted out of shape if the *full* centerline is not constructed for the circle. The minor axis will remain parallel to the receding axis of the cylinder or hole, as is shown in Figure 2-2. This receding axis is often referred to as a *thrust line*.

No other features pop out more when incorrectly drawn than do holes, cylinders, or cir-

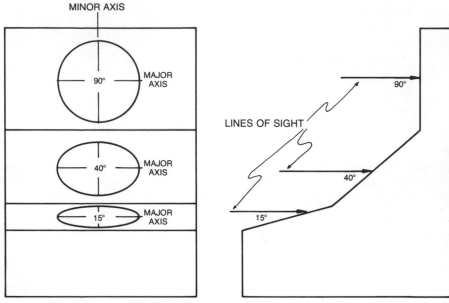

Figure 2-1. Ellipse exposure.

Figure 2-2. Minor axis aligned with thrust axis.

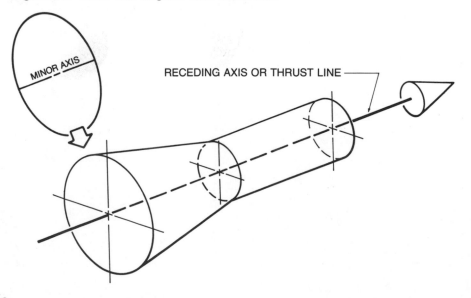

cular objects. Study Figure 2-3 to see how ellipses are oriented in relation to centerlines and thrust lines.

The illustrator must choose the correct template, size it, and physically move it across the drawing until it is in the desired position. Computers used as illustration tools also have templates that the illustrator can select and position on the drawing. In this case the template is stored in electronic memory, is called up from a menu, and is moved around the graphics screen by a positioning device. The function of an electronic template, such as that shown in Figure 2-4, is the same as for plastic templates. The same decisions have to be made with both.

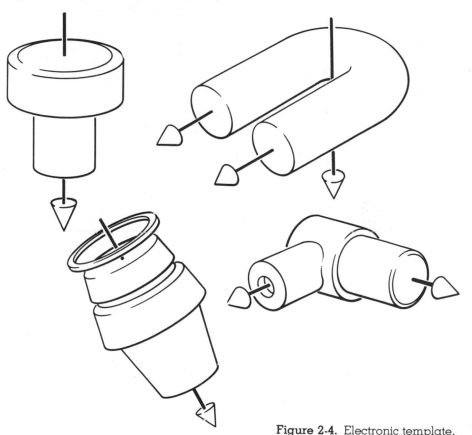

Figure 2-3. Examples of axis alignment.

Figure 2-4. Electronic template.

ROT AZ	TRIM	*VER*		*CHG*	
MIR	STR	*NTXT*		*NFIG*	
MOV COPY	COPY ENT TO	TXT	PNT	FIG	STG
MEAS DIST	INTOF	PRT	LDIM	MX	STCH
MOD NLIN	SET HIDN	PNT XOYO	XH	MY	SCL
DIP14	CAPD	PC.BUL	DIODE	RES	AND
DIP16	CAP	ARS	NPN	RESV	NAND
DIP8	CAP+	VCC	PNP	CAP	OR
RES4	DIODE	GND	IN	CAPV	-OR
RES8	TRANS	SCH.BUL	OUT	CAP+	INV

Construction on Grids

By using scale underlay grids, as shown in Figure 2-5, you can reduce layout and construction time. The blocks or divisions can be rescaled on the overlay, with decimal or fractional divisions made as needed. Once you become skilled in using grids, you can draw parallel lines without resetting your drafting machine head or T-square.

Using a grid also helps you visualize three-dimensional space on paper. Without the grid, the paper is a flat plane. With the grid, the space is better defined and you can see three-dimensional form and parts placement.

It is not important to start your drawing at the corner of the grid box. In fact, if you start there, your drawing may obscure the scales, making the drawing more difficult. Notice in Figure 2-5 that the construction for an object has been started away from the corner, allowing measurements to be picked off without looking through the construction.

Several axonometric grids have been included in this text for your use.

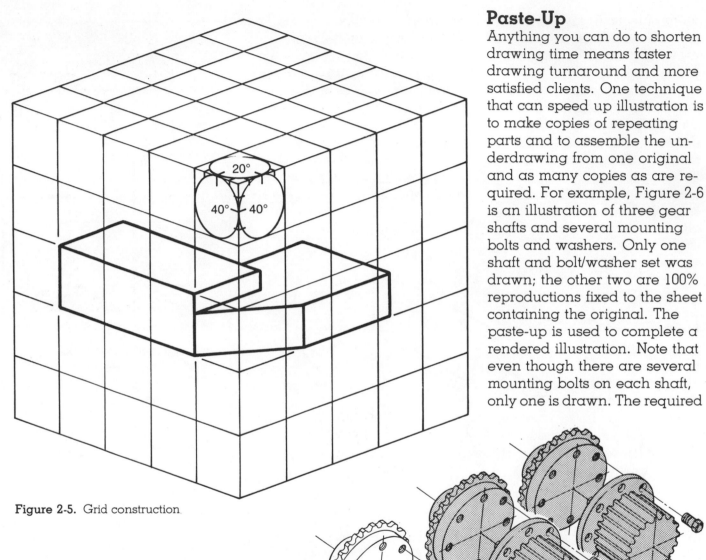

Figure 2-5. Grid construction.

Paste-Up

Anything you can do to shorten drawing time means faster drawing turnaround and more satisfied clients. One technique that can speed up illustration is to make copies of repeating parts and to assemble the underdrawing from one original and as many copies as are required. For example, Figure 2-6 is an illustration of three gear shafts and several mounting bolts and washers. Only one shaft and bolt/washer set was drawn; the other two are 100% reproductions fixed to the sheet containing the original. The paste-up is used to complete a rendered illustration. Note that even though there are several mounting bolts on each shaft, only one is drawn. The required

Figure 2-6. Paste-up.

number of bolts is recorded on the appropriate parts list.

Paste-ups can also be done from a photographic library of standard parts. If a company uses standard parts and settles on a standard axonometric view, drawings of parts can be shot at different percentages for different scales.

The more illustration you do, the greater the opportunity you will have to use these paste-up shortcuts.

Mechanical and Electronic Aids

Companies and individuals are continually introducing aids for producing faster and more accurate illustration. Some of these aids work well, whereas others only make up for a lack of skill on the part of the illustrator and may even keep the illustrator from learning the basics.

Most supervisors and illustrators will tell you that the more you know about illustration, the more you can use these devices. The less you know, the more devices will use you.

These aids fall into three general areas: (1) mechanical aids, (2) photo mechanical aids, and (3) electronic aids.

MECHANICAL AIDS

Axonometric scales (Figure 2-7) are produced for several standard views. Also available are axonometric triangles, in which the angles correspond to the axis angles. Scales and triangles can be combined into one unit and affixed to a drafting machine by a standard chuck. This sort of aid can be a powerful tool in direct axonometric construction.

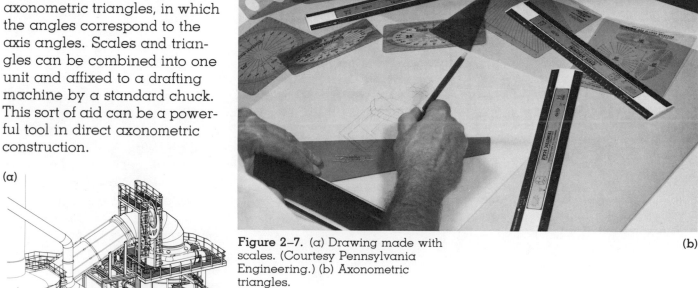

(b)

(a)

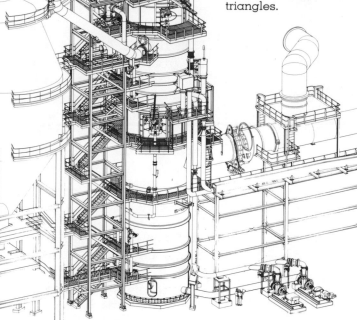

Figure 2–7. (a) Drawing made with scales. (Courtesy Pennsylvania Engineering.) (b) Axonometric triangles.

Perspective and axonometric tables (Figure 2-8a) allow drawings to be made without normal set-up and construction. An axonometric table is effective and flexible because it can be used to produce an infinite number of axonometric views. An axograph, such as the one pictured in Figure 2-8a, provides a wide range of views.

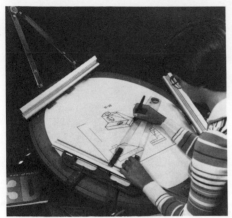

(a)

(b)

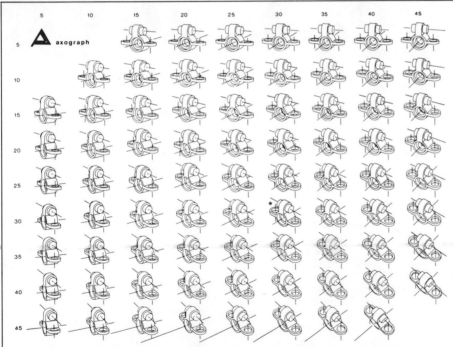

Figure 2–8. (a) Axograph tables and (b) axograph views. (Courtesy of Axo-graph Ltd.)

PHOTOMECHANICAL AIDS

Taking a photograph of a drawing at an angle will produce a twisted or axonometric image. Unless the angle is great, or the object quite long, the foreshortening will be nearly constant. Architects who can't or won't learn drawing often use such photographs to produce accurate *perspective* drawings. However, as much time and effort can be expended in this method of trying to avoid construction drawing as would be expended in doing the drawing itself. For this reason photomechanical techniques should supplement direct construction and not be used in lieu of it.

ELECTRONIC AIDS

Electronic illustrations have been around for some time. Their expense has precluded all but a few from using them, but as more and more industries have become computerized—have developed high-speed digital computing capability—the chance for computer illustration being available to most practitioners has greatly increased.

Early machines, such as the Illustromat, were of the trace-over variety. Their accuracy

Figure 2–9. Illustromat. (Courtesy of *Graphics Today.*)

depended on the skill of the operator, who had to physically determine line visibility by the movement of the tracing stylus (Figure 2-9).

Today, many computer graphics companies have included illustration programs as part of their graphics capabilities. The image still serves as an underlay, but now it can be stored, recalled, and altered without having to retrace the detail drawings (Figure 2-10). In addition, these graphics devices can talk to other computer-aided devices, such as typesetters or word processors, making for direct image/text composition.

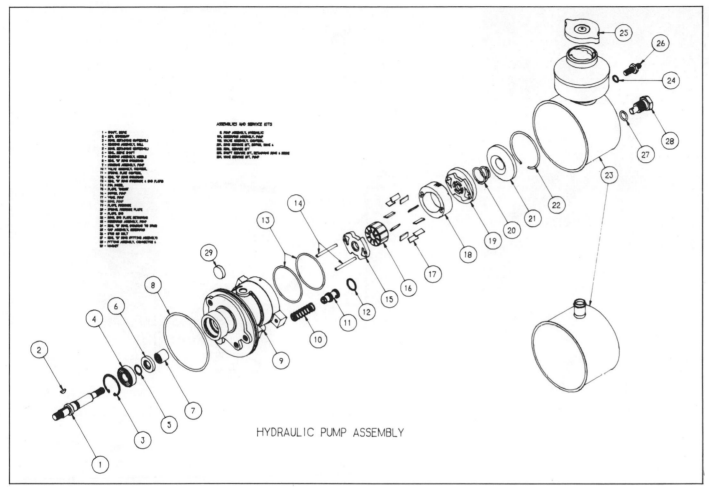

Figure 2-10. Computer illustration of hydraulic pump assembly.

Photographic Underlays

Photographs can be used as underlays for drawings so that you do not have to construct original pictorial drawings. The image will be perspective rather than axonometric, especially when wide-angle lenses are used. Photographs are used when:

1. No detail drawings exist for the subject, as might be the case with prototypes, defunct or uncooperative suppliers, and so on.

2. Actual installation is the product of technician decisions, as would be the case in electrical or hydraulic line routing. The engineering drawings or schematics only show design relationships—the actual placement of components is done on the shop floor.

3. What you are trying to illustrate may exist for only a short time before it is painted, welded, or covered by other components.

To use this method, a fine-grain print must be made of the subject at a scale large enough for you to make an accurate tracing. Both film and vellum work well over photographs. Tape the photograph down to your

 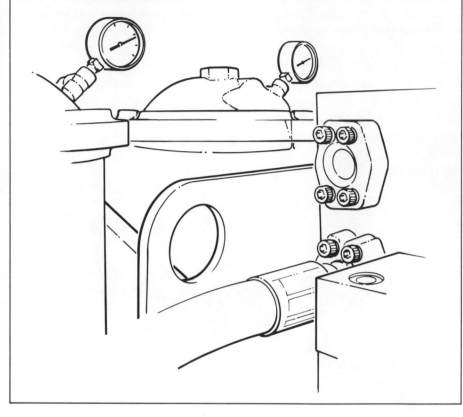

Figure 2-11. Photograph and photodrawing.

work surface. Tape the overlay material securely over the photograph, using one long strip of tape across the top to act as a hinge. Use one or two pieces of tape at the bottom so that you can easily lift the overlay whenever necessary to study the photograph.

Line-weight rendering is a fast and effective method of treating subjects traced from photographs (Figure 2-11). Here's a word of advice: Start at the center of what you are trying to show, using good detail and lots of contrast. But don't make the mistake of trying to show too much. Allow the illustration to blend out from this central area in a *vignette*.

Computer Underlays

Computer illustration, as shown in Figure 2-12, can replace the constructed layout *if* much of the illustration is highly repetitive. But if every object is different and must be geometrically defined from scratch, a topnotch illustrator can compete with the computer. Furthermore, the computer drawing can serve only as a guide for the hand-drawn finished illustration.

Figure 2-12. Computer drawing of pump housing and shaft assembly.

Nevertheless, tremendous advances have been made recently in areas of illustration where manual illustration has been previously more efficient: hidden feature removal, changing line weights, casting shadows, leaving highlights, and so on. These advances represent hundreds of man-years of development time. (Imagine how much illustration could

have been turned out by illustrators in the same time!)

The bottom line in computer drawing is the cost per unit of illustration. It takes a sizable volume of illustration as well as a large amount of money to

invest in a computer system. And skilled illustrators are still needed to turn the computer drawings into camera-ready illustration.

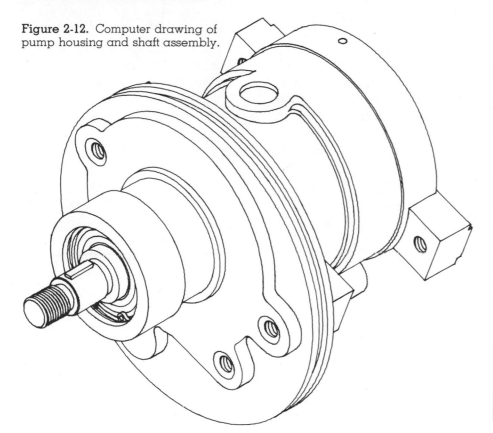

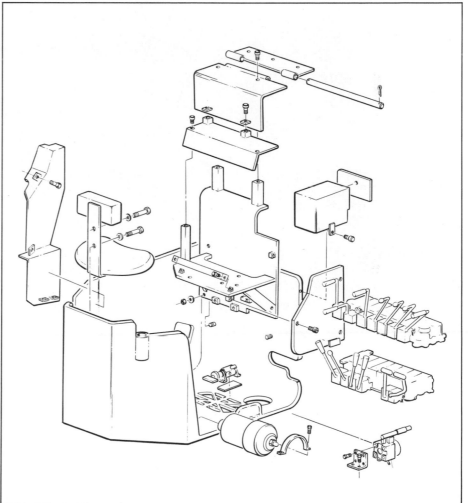

Figure 2-13. Camera-ready illustration should be clean and clear. Lines should be bold but with contrast between outline and edge line.

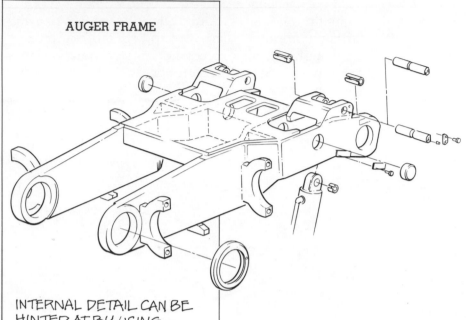

AUGER FRAME

INTERNAL DETAIL CAN BE
HINTED AT BY USING
DASHED LINES, WHICH
GIVE A GOOD IMPRESSION
OF THE HIDDEN DETAIL.
PARTS CAN BE ALLOWED
TO FADE AWAY
(VIGNETTE), AS IN THE
CASE OF THE HYDRAULIC
CYLINDER, SHOWN IN
FIGURE 2-11, IF THEIR
FULL DETAIL IS NOT
IMPORTANT TO THE
ILLUSTRATION.

Basic Tools and Practices of Technical Illustration

The tools of technical illustration range from simple to complex and cost from a few cents to several hundred thousand dollars. They include hardware and software, expendables and consumables. These tools are an integral part of the basic practices of technical illustration, so this section will describe both tools *and* practices.

CLASSIFYING THE TOOLS

All of the tools discussed in this section have been assigned a code representing their business classification. Such classification is particularly useful to freelancers—and all illustrators at some time or another work some freelance jobs. The classification scheme is:

Capital Equipment (CE). This category includes hardware, furniture, and equipment costing over $100 that has an extendable usable life. Timely purchases of capital equipment are to the benefit of companies, agencies, and individuals. Some supplies that would otherwise be direct expenditures—usually as operating expenses—can also be considered capital purchases. Illustrators should purchase *only* the capital equipment they need to make money and should purchase the services of occasionally needed equipment. Because used capital equipment can still be depreciated and tax credits can be taken on it, buying used equipment is often the best investment over the long run.

Expendable Equipment (EE). The technical illustrator requires a considerable amount of equipment that is treated as consumable even though it has a usable life. Scales, guides, instruments, and the like are not usually depreciated unless they are expensive. If properly cared for, these items last a lifetime. If mistreated, they must be purchased again and again.

Expendable Supplies (ES). An amazing amount of material is consumed in the production of technical illustration. The cost of such materials, combined with the services that at times have to be purchased, represents a considerable outlay. These supplies and services are a direct cost of doing business and may be treated as such. A final note: other businesses' expendables may become *your* capital equipment. Look for useful discards that you can acquire for little or no cost.

LISTING OF TOOLS AND PRACTICES

Acetate (ES). Cellulose acetate is a clear plastic film that is used in several ways. It comes in sheets approximately 25" x 40" or in rolls 20" or 40" wide by 50' to 100' long. Thickness determines what the acetate can be used for. Thicknesses of .003" to .020" are common. Acetate may be purchased with both sides slick or with one side frosted to aid in taking ink, which would normally be repelled by the slick surface. Color-tinted acetate is available for use as an overlay material for separations or as a protective covering. Photo-transfer techniques can be used for directly producing line illustration on clear film, thus reducing the time required to ink on acetate. The resulting *cels*, as they are

called, may be back-colored with paint or dye or may be used in combination with existing illustration. When acetate is needed solely as a protective covering, cellophane may be substituted at a lower cost.

Adhesives (ES). The adhesives used in technical illustration are primarily of the contact variety, rather than wet gluing. Adhesive methods range from dry mounting, waxing, cementing, and bonding to simple taping. The old stand-by, of course, is rubber cement—a form of lightweight contact cement. The difficulty in spreading rubber cement over large areas has been solved by use of spray adhesives. Spray adhesives are clear and can bond almost any porous, semiporous, or nonporous material. The least permanent adhesive is wax, applied with either a hand roller or a sheet waxer. Wax remains the common adhesive for paste-up because waxed materials can be removed and repositioned.

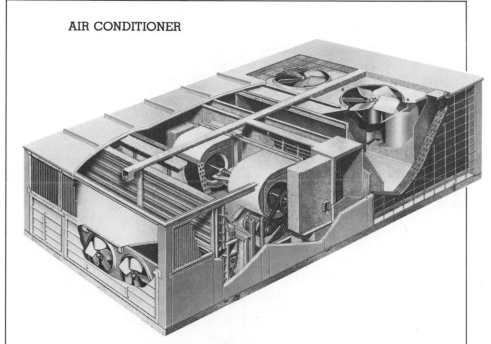

AIR CONDITIONER

IF AN ILLUSTRATOR WORKS AS AN EMPLOYEE OF A COMPANY, ALL HIS OR HER ILLUSTRATION IS USUALLY THE PROPERTY OF THAT COMPANY. THE ILLUSTRATION IS USUALLY THE PROPERTY OF THAT COMPANY. THE ILLUSTRATION CAN BE USED OVER AND OVER AGAIN FOR DIFFERENT PURPOSES — IT CAN BE COPIED, FRAMED, PUBLISHED, PHOTOGRAPHED — WITHOUT FURTHER COMPENSATION TO THE ILLUSTRATOR. THIS IS THE PRICE THE EMPLOYEE PAYS FOR THE SECURITY OF WORKING FOR A COMPANY.

ILLUSTRATORS WHO WORK FREELANCE MAY GIVE UP CONTROL OF THEIR WORK IF THEY SEE THAT THE ILLUSTRATION IS OF USE ONLY TO THE PARTICULAR CLIENT, AS WITH THIS AIR CONDITIONER. BUT IF THE WORK HAS INTRINSIC ARTISTIC VALUE OR IF IT CAN BE USED AGAIN BY THE ILLUSTRATOR ON SUBSEQUENT JOBS, HE OR SHE MAY WISH TO COPYRIGHT IT. THE CLIENT IS GRANTED UNLIMITED USE OF THE ILLUSTRATION TO PROMOTE THE CLIENT'S PRODUCT, WHILE THE ILLUSTRATOR KEEPS CONTROL OF THE ORIGINAL. IN THIS WAY THE CLIENT DOESN'T END UP WITH ANY LESS, WHEREAS THE ILLUSTRATOR RETAINS THE RIGHT TO PROFIT AT A LATER DATE FROM THE ILLUSTRATION. THE ILLUSTRATOR WHO WISHES TO RETAIN SUCH RIGHTS NEEDS TO MAKE SURE THAT A LEGAL DOCUMENT IS DRAWN UP AND SIGNED BY ALL PARTIES.

CLIENTS DO HAVE PROPRIETARY RIGHTS THAT MAY PRECLUDE AN ILLUSTRATOR FROM KEEPING CONTROL OF THE ILLUSTRATION. THIS IS ESPECIALLY TRUE WITH PRODUCT DEVELOPMENT OR RESEARCH. IN SUCH CASES THE ILLUSTRATOR MAY BE REQUIRED TO RETURN ALL MATERIALS (PHOTOS, ENGINEERING DRAWINGS, SKETCHES, PRINTS, AND UNDERLAYS) TO THE CLIENT.
(COURTESY OF LENNOX INDUSTRIES INC.).

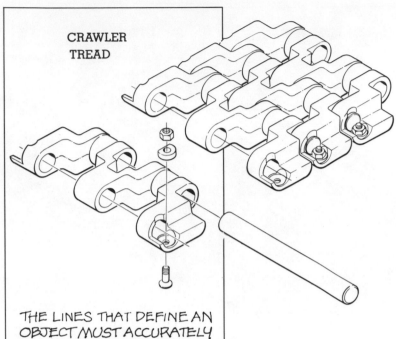

CRAWLER
TREAD

THE LINES THAT DEFINE AN OBJECT MUST ACCURATELY TELL THE VIEWER THE SHAPE AND MASS OF THE OBJECT. OUTSIDE LINES ARE THE MOST IMPORTANT BECAUSE THEY SET THE OBJECT'S LIMITS. EDGES SHOW CONTOUR AND PLANE CHANGE. FLOW LINES SHOW DIRECTION OF ASSEMBLY. TOGETHER, THESE LINES MAKE UP EFFECTIVE TECHNICAL ILLUSTRATION.

Airbrush (CE or EE). The airbrush (Figure 2-14) is a precision instrument that delivers a colorant atomized in a controlled stream of air. Quality airbrushes control the stream of air and the amount of colorant exposed to the stream. The airbrush is supplied with air from a compressed air tank, from a holding tank connected to a compressor, or from an aerosol canister.

An airbrush can be used in several distinct ways. First, it may be the only technique used to produce an illustration. Fully airbrushed illustration is expensive because of the amount of masking and frisketing required. Second, it may be used for photo-retouching. Photo-retouching is not the important technique it once was, because photography has improved, printing processes have become more flexible, and drawn illustration is used more than

photography. Finally, the airbrush is useful for applying large areas of tone to backgrounds and other areas. It is likely that photo-retouching will continue to lose importance, as will the totally airbrushed illustration.

Axonometric Projection. This type of pictorial drawing virtually distinguishes the technical from the commercial illustrator, since commercial illustrators may never need to use axonometric projection or drawing. Axonometric projection allows linear scaling along axes, making the exact relationship between features measurable. There are three types of axonometric projection: isometric, the least realistic; dimetric; and trimetric, the most realistic (Figure 2-15). Trimetric projection is

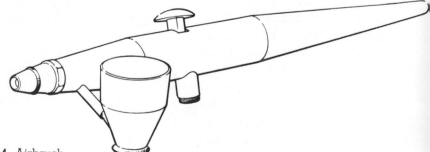

Figure 2-14. Airbrush.

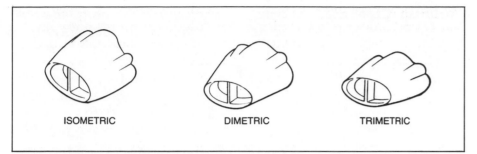

ISOMETRIC   DIMETRIC   TRIMETRIC

Figure 2-15. Axonometric projections.

the most like perspective drawing yet retains an ease of measurement, without perspective's progressive foreshortening (see Chapter Five). Several of the most common axonometric views are available with commercial guides and scales. Furthermore the advent of computer illustration has opened the door for rapid axonometric and perspective drawing without constructing special scales.

Blueprinting (ES). Most technical drawing needs to be reproduced without damage to the valuable original. Thus, many technical drawings are not kept or handled in their original form but rather are kept photographically or electronically and are reproduced by making a print. The term *blueprint* is used to describe most engineering prints, even those produced by other processes. In addition to blueprints there are more stable intermediate prints from which further prints can be made. These are often called "sepias" because of their reddish-brown line color.

Burnisher (ES). The burnisher (Figure 2-16) is a basic tool used in layout and in the preparation of technical illustration for reproduction. It is used to position paste-up items and to press down—or *burnish*—waxed type, presstype, pressure-sensitive film, and so on. If a commercial burnisher is not available, any smooth object,

Figure 2-16. Burnisher.

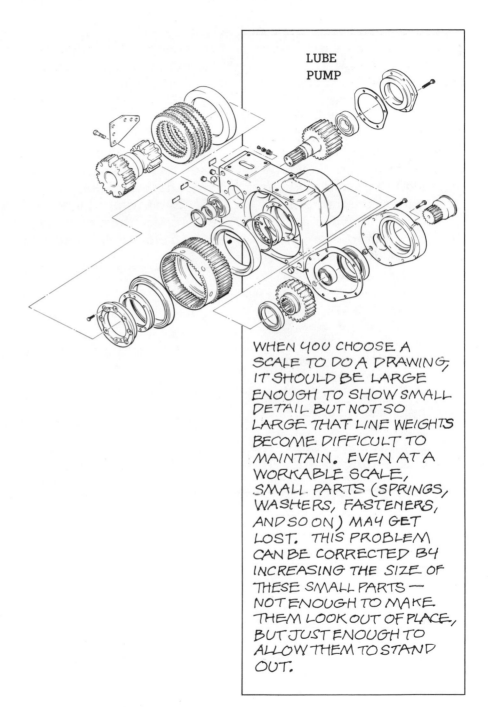

LUBE
PUMP

WHEN YOU CHOOSE A SCALE TO DO A DRAWING, IT SHOULD BE LARGE ENOUGH TO SHOW SMALL DETAIL BUT NOT SO LARGE THAT LINE WEIGHTS BECOME DIFFICULT TO MAINTAIN. EVEN AT A WORKABLE SCALE, SMALL PARTS (SPRINGS, WASHERS, FASTENERS, AND SO ON) MAY GET LOST. THIS PROBLEM CAN BE CORRECTED BY INCREASING THE SIZE OF THESE SMALL PARTS — NOT ENOUGH TO MAKE THEM LOOK OUT OF PLACE, BUT JUST ENOUGH TO ALLOW THEM TO STAND OUT.

such as the end of a ballpoint pen cap, can be used for burnishing.

Cameras (CE). Technical illustrators may need to use several cameras in making illustration. The lucidagraph, or *camera lucida*, is used to enlarge or reduce an image to be traced. If illustrators have access to a comprehensive photo lab, one with a stationary or "stat" camera, they can do enlargement or reduction without redrawing. However, the camera lucida does have one advantage over the stat camera: it can be used to project three-dimensional set-ups, which can then be traced. Many technical illustrators find that they need to be accomplished technical photographers, too.

Cathode Ray Tube (CE). The CRT is emerging as the technical illustrator's sketch pad. Hooked up to a computer, it can be equipped to refresh the image, giving the illusion of motion, and it can be made sensitive to external commands through touching or keying. The ability to draw directly with the computer is called *interactive computer-aided graphics.*

Color Systems. There are several systems of specifying color, but all rely on three components: *hue, chroma,* and *value* (Figure 2-17). The main systems used today are those by Ostwald, Munsell, and Hickethier. Color systems can be traced to Athanasius Kirchner's two-dimensional color chart of 1671. Systems used today are color *solids.* The *hue* is the particular color as distinguished from all other colors. For example, a particular yellow and a particular green would be different hues, and one hue of yellow would be different from all other hues of yellow. The *chroma* in a specific hue determines the saturation or intensity of the color. The *value* associated with the hue and chroma is its relative lightness or darkness, irrespective of the hue or chroma. The importance of value can be seen if you make a color illustration using different hues and vary the chroma within those hues but keep all the colors at the same value (reflectivity), then take a black-and-white photograph of the illustration. With all values the same, little of the original detail will show up in the photograph. The most important aspect of color systems for the

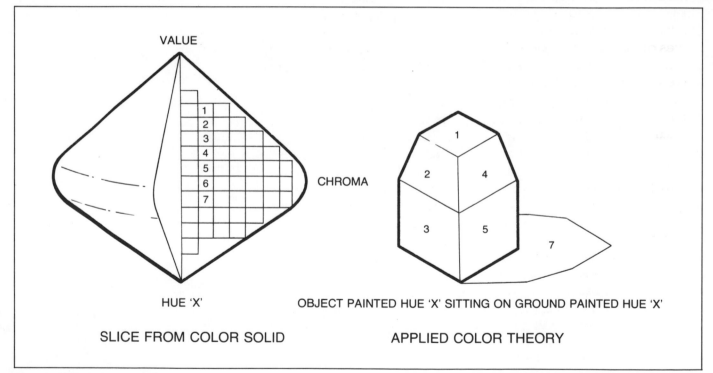

VALUE

CHROMA

HUE 'X'

SLICE FROM COLOR SOLID

OBJECT PAINTED HUE 'X' SITTING ON GROUND PAINTED HUE 'X'

APPLIED COLOR THEORY

Figure 2-17. Color systems.

technical illustrator is to internalize a system so that the illustration is consistent from one part of the drawing to another and consistent from one illustration to another.

Curves (ES). The technical illustrator uses curves in a way different from other artists. In technical illustration, complex data points must be plotted, prediction curves need to be drawn, and contours must be plotted and then smoothed. Most curves are made of clear plastic. The highest quality curves are die cut from extruded acrylic sheet, whereas cheaper curves are injection molded and are susceptible to chipping and cracking. Curves are either regular, as in highway curves, or irregular, as with French curves. Adjustable curves come in many forms: rubber built around a lead core, kerfed plastic, or a clear plastic edge held by weighted splines. They are good for large, sweeping curves but are less effective for small, tight curves.

Drafting Machine (EE or CE). Drafting machines increase the accuracy and efficiency of technical illustration by combining the functions of T-square, protractor, scale, and straight edge. There are several sizes and grades of two designs: (1) the parallel arm machine, which is portable and adaptable, and (2) the track machine, which is more permanent and often more accurate. The technical illustrator uses a drafting machine to produce accurate technical drawings and to produce rapid, accurate keylines and paste-up.

Drawing Table (CE). The surface on which the technical illustrator does the drawing must be blemish-free. This situation is often difficult to maintain because of the different functions required to produce illustration. Thus, many artists have found they need three pieces of furniture:

1. A *drafting table* for technical drawing, tracing, keylining, paste-up, and inking.
2. A *drawing table* for wet illustration, airbrush, model building, and reference.
3. A *cutting table* for stripping, cutting, and applying adhesives.

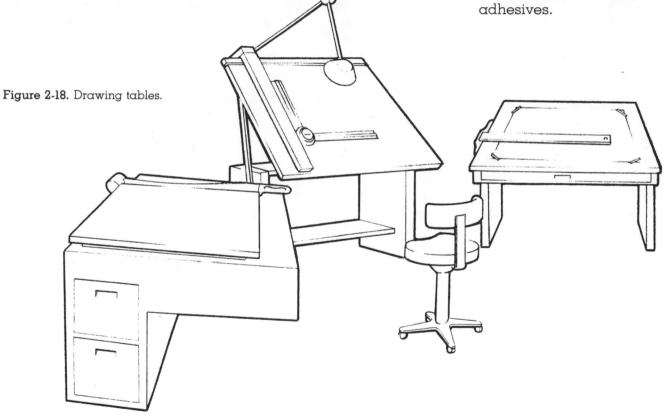

Figure 2-18. Drawing tables.

Dry Mounting (ES). Dry mounting is often considered to be the best way to mount artwork, drawings, and photographs. A dry mount press is used to heat the adhesive tissue and hold the art and backing board under pressure until they are bonded. However, dry mounting is not a permanent method of mounting, nor is it advisable for translucent originals. Thus, in many cases dry mounting has been replaced by the use of spray adhesives.

Dyes (ES). Dyes are liquid, water-soluble colorants sold in small bottles, either individually or in sets. They offer brilliant watercolor technique, with fine pigment carried evenly in suspension. Dyes should not clog the finest drawing instruments. Thus, they are usually used when a colorant containing a ground or binder would clog an instrument. Dyes require great skill from the illustrator—they can be very unforgiving in their bleeding, staining, and saturating.

Electric Eraser (EE). Electric erasers make it easier to change major parts of a drawing. However, if a *very* large part of the drawing needs to be changed, the fastest route may be cutting and pasting an intermediate.

An *erasing shield* should be used to control the rather large eraser stock. The technical illustrator uses the electric eraser in two ways: (1) to erase parts of an existing drawing that are incorrect, out of date, or unnecessary, and (2) to produce desired effects in a piece of illustration. It is often easier to airbrush an area and erase out the detail than to mask and frisket.

The technical illustrator needs to know how the drawing will be handled after it leaves the board. If it is to be photographed, certain things can be done to the art that would be inappropriate if the illustration were to be displayed as an original. For example, parts of the drawing can be temporarily covered with white paper, something that would be noticeable to the eye but ignored by the camera. "White out" or white opaque can be used to cover up small mistakes.

Ellipse Guides (EE). Ellipse guides are among the many types of guides that make up an entire system of templates available to the technical illustrator. The ellipse guide allows non-normal views of circles to be drawn more accurately.

There are four sizes of ellipse guides:

1. Giant ellipses over 4″ in diameter.
2. Large ellipses over 1½″ in diameter.
3. Small ellipses ⅛″ to 1½″ in diameter.
4. Very small ellipses under ⅛″ in diameter.

There are several grades of templates, the highest being die cut from extruded plastic sheet. Guides that are cut by a router can present irregular edges, so they should be inspected before being used. Do not stick tabs, coins, or tape to the underside of the guides to elevate them for inking. Instead, use guides in combination to avoid smearing linework.

Erasers (ES). Having the correct eraser for the job is important to the technical illustrator. If the grit of the eraser is too fine, much time will be wasted. If the grit of the eraser is too coarse, the illustration may be damaged.

Erasers may be wet or dry; they may be for ink, for graphite, or for both:

1. *Liquid eraser* is used on intermediates and film; it dissolves rather than erasing in the usual manner.

2. *Graphite erasers* include pink pearl, kneaded, art gum, plastic, and green nile.
3. *Ink erasers* include ruby red and other high-abrasive erasers, as well as eradicator-impregnated plastic erasers.
4. *Plastic erasers* not containing abrasive or eradicator are used on paper or drawing film.

Filing Systems (CE). By far the most popular filing system for flat illustration is the *flat file* or *plan chest*. This file keeps drawings flat and away from light and dust. The best files have ball-bearing drawer rails. Older files made of hardwood are just as durable as steel files. Used filing cabinets, if they can be found, offer an alternative to new equipment. If illustration is formatted for books, vertical legal files may do well. The drawback with vertical filing is that flimsy tracings must be mounted in order to prevent damage. Large drawings and illustrations can be stored rolled up, but the sheets become unmanageable when unrolled. If you are working with drawing sets, they can be hung from racks.

The most efficient means for filing is using microfilm aperture cards or direct electronic storage. In both cases, vast amounts of information can be stored in a small space. Back-up systems should be provided to guard against accidental damage or erasure.

Three-dimensional illustration (models, prototypes, and mock-ups) pose a different problem. They are often so large that to be moved they must be sectioned and reassembled. Smaller models may be enclosed in glass or plexiglas cases to keep dust, dirt, and handling damage to a minimum. If neither of these approaches is feasible, detailed photographs should be taken before the model is destroyed.

Fixatives (ES). When technical illustrators must protect surfaces from soilage beyond using a cover flap and tissue overlay, they make use of fixatives. Fixative also comes in handy when one layer of pigment must be kept from bleeding through to subsequent applications. In addition, fixatives are used to protect photographs, cover presstype, and seal type proofs before pasteup. Fixatives are of two varieties:

1. *Nonworkable fixatives*, which provide a fairly slick surface to which paint, ink, dyes, and pencil will not adhere.
2. *Workable fixatives*, which provide a matte finish with sufficient tooth to accept additional applications of medium.

No matter which type of fixative is used, a slow, even application will not discolor, bleed, or build up.

Frisket (ES). Frisket is a product used to protect parts of an illustration while working on other parts of the drawing. Using frisket, the entire surface of an illustration may be done as a series of windows through which areas are rendered. Frisket is available in three forms:

1. *Nongummed frisket*, a thin tissue to which adhesive must be applied.
2. *Pregummed frisket*, a tissue or film with a light adhesive applied to one side. A backing sheet is used to separate sheets.
3. *Liquid frisket*, which is much like thinned rubber cement (many airbrush artists actually use rubber cement as a frisket) with a light pigmentation so that covered parts of the drawing may be readily identified.

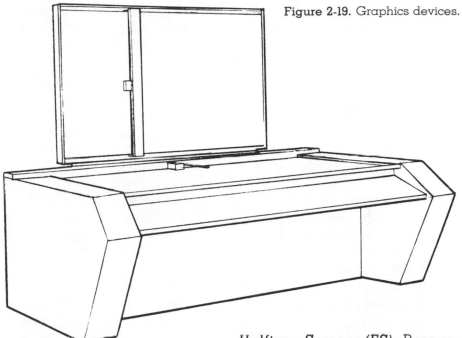

Figure 2-19. Graphics devices.

Graphics Devices (CE). It is becoming more common for technical illustrators to use data or images from a computer in their work. The illustrator can personally generate the drawing or can use images made by computer operators. These images may be:

1. Output from a drum or flatbed plotter representing the set of points in memory.
2. Hard-copy output, either photographic or dry copy from an on-line CRT.
3. Strip charts or plots from a strip printer.

Halftone Screens (ES). Because of the limitations of the offset lithographic printing process, continuous tone art cannot be reproduced directly. Even photographs, which are not actually continuous in their tones, cannot be directly printed. The method of translating a photograph or continuous tone drawing into a form that is printable is called the *halftone*, in which the image is photographed through a fine screen, creating a pattern of minute dots. Similar to this is the *Ben Day* process, which mechanically, rather than photographically, converts continuous tone into a pattern of dots.

Figure 2-20. Handrests.

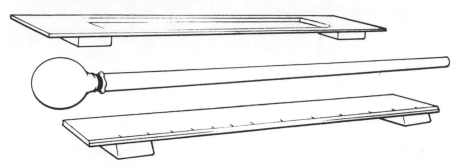

Handrests (Bridges) (ES).

Handrests are used when an illustrator has to work over areas that are wet with paint or ink or that would be soiled if touched. Three types of rests are available (Figure 2-20):

1. The mahlstick, which is a dowel with a soft chamois pad on the end. It is generally used to support the hand or arm for work on vertical surfaces.
2. Commercial rests, which serve as combined handrest, equipment tray, and ruling edge.
3. Homemade rests, which generally serve as straight edges and can be made from scrap materials.

Illustration Board (ES).

"Illustration board" is a general term referring to a large group of boards and heavy papers. True illustration board is quality drawing paper bonded to an inexpensive backing board. Certain boards and board surfaces work best with various media, and only through experience can an illustrator identify the best boards for particular purposes. Board may be purchased in various sizes trimmed from 40" x 60" and in varying thickness of both facing paper and backing board. Types of illustration board include:

1. Hot press board, which has a smooth plate finish and tends to repel many paints and dyes.
2. Cold press board, which has a slightly grained finish.
3. Rough finish board, which is more suitable for watercolor or pastel.

Technical illustration should be done on board in order to protect the art. In cases where the illustration cannot be done on board, as with tracing, fine linework, or front and back rendering, the finished illustration should be mounted to a board.

Because mounting an illustration increases the amount of space required to store it, some companies may prefer that illustrations not be mounted. In these cases, illustrations should be lightly taped to a board with a cover flap. The clients can then remove the illustrations as they desire.

Inks (ES).

The introduction of the technical inking pen has drawn attention to the limitations of inks. Inks that flow well in drawing pens may not cover well, and inks that cover well may not flow easily. For example, some inks have pigment so fine and so uniform in suspension that they will not clog a 6X0 pen. However, these inks may not be opaque. The trade-off between coverage and flexibility has been solved by at least one company, but many inks still represent a compromise between coverage and ability to flow.

Keylining.

Keylines are the lines that the technical illustrator uses to organize an area in terms of type placement, photograph location, art position, and so on. If they remain on the art or cover tissue, they can be ignored by the photographic film or stripped or opaqued on the negative. Nonreproducible blue pencil is generally used for keylining.

Knives (ES).

The knives most commonly used by the illustrator are the matte knife and the trimming knife (Figure 2-21). The matte knife is used for cutting illustration board and matte board. Although there are guides and aids available for

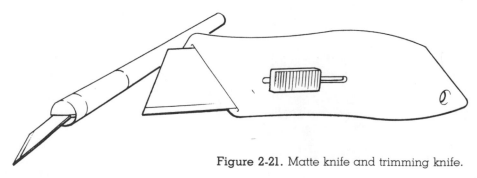

Figure 2-21. Matte knife and trimming knife.

Figure 2-22. Lead holder.

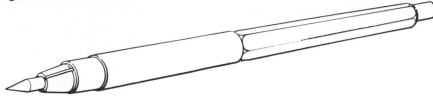

these cutting jobs, none is a substitute for skill and dexterity, as the novice illustrator will soon find out. The *trimming knife* is used for stripping type, negatives, frisket tissue, and paper. Trimming knives have replaceable blades in a choice of styles, and some knives are designed to pivot and swivel.

Lead Holder (ES).   Because much of a wooden pencil is wasted in the sharpening process, many illustrators use a common holder for stick graphite (Figure 2-22). The graphite can be used down to

½", yet the holder remains a comfortable length. There is no reason why a lead holder, if properly cared for, cannot last several years—or even as long as an entire career. The gripper feet in the chuck should be periodically tightened by rotating the feet with the plunger depressed. While most holders are designed for 0.78" lead, some also accommodate. 05" or .01" lead.

Lead Pointers (EE).   Several styles of lead pointer are available (Figure 2-23), including

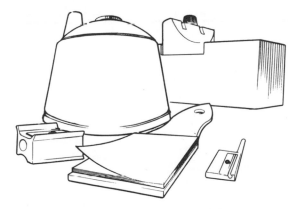

Figure 2-23. Lead pointers (sharpeners).

ones that can sharpen different diameters of lead. Generally speaking, the less expensive models do not work as well over the long run. And there is nothing more frustrating than a faulty lead pointer.

Pointers work either by abrading the graphite with abrasive paper or by cutting the graphite with metal blades.

Leads (ES).   Drawing leads are not, of course, lead at all. Rather, they are cylinders of compressed graphite held together by a binder. The less binder, the darker the line produced. The greater the percentage of binder, the harder the lead and the lighter the line. The technical illustrator knows the limitations of each grade of lead and knows how each grade "takes" to a particular surface. Plastic leads are available for use on film, and leads can also be purchased in a variety of colors. Common gradations of both graphite and plastic leads are shown in Table 2-1.

Table 2.1. Drawing Leads

	Graphite	Plastic
Soft	2B-B	E1
Medium	HB-F	E2, E3
Hard	H-4H	E4
Very hard	5H-9H	E5

Figure 2-24. Leroy lettering.

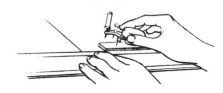

Leroy Lettering (EE or ES).   The technical illustrator must be able to rapidly produce supporting text, either by mechanical devices, by hand, or by lettering template. Leroy lettering is a mechanical pantographic method of producing consistent letters one at a time (Figure 2-24). There are many brands besides K & E Leroy but all operate essentially in the same manner. There are many mechanical typefaces available,the most common being a highly readable yet unexciting technical engineering face. Technical fountain pens now universally fit into the mechanical scriber, making constant refilling of the old-style penpoint unnecessary. Proficient illustrators can letter in this manner almost as fast as they can by hand.

Figure 2-25. Markers come in many forms.

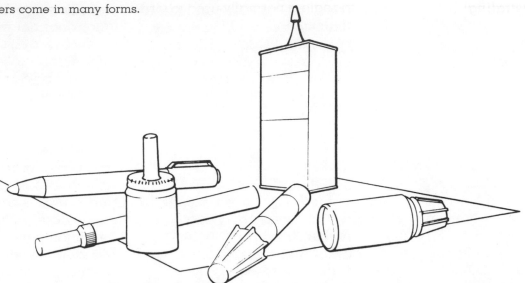

Lettering Templates (ES). The lettering template is another mechanical lettering device, although it is not as accurate as the Leroy-type system. With practice, an illustrator can achieve credible results. Lettering templates offer few typefaces and usually have router-cut openings.

Another type of lettering template is that used by computer typesetting machines and graphics terminals. Unlike older typesetting equipment, which uses physical fonts, wheels, strips, or the like, these machines have electronic fonts, giving greater flexibility in copy-fitting, as well as the opportunity to tie in with other

computers. The images in electronic memory may be thought of as templates moved around the drawing by the operator.

Light Table (CE). A light table has a translucent surface illuminated by lights below, thus allowing for accurate tracing. It also provides backlighting for opaquing and stripping negatives, mounting slides, and overlaying mechanicals. A light table can easily be built for a fraction of the cost of a commercial unit.

Markers (ES). Markers are the most flexible product for quick sketching. In skilled hands, they can be used like brushes. It is important to remember that markers have certain capabilities and certain limitations.

They should not be expected to do things they cannot. Markers are either permanent or non-permanent. Some of the oil-based markers that are considered permanent really are not—as some illustrators have learned the hard way. Markers come in different shapes and sizes (Figure 2-25), the popular studio markers may be clumsy for large hands and the longer barrel markers may be unwieldly for smaller hands. Some of the least expensive markers turn out to be the best. Don't throw away markers! An old dry marker may be able to produce the effect you are looking for—an effect a new, wet

marker would be incapable of producing.

Mechanicals (ES). The preparation of camera-ready illustration is a basic function of the technical illustrator. Camera-ready art is called a "mechanical" because the printing processes required to produce the art are mechanically separated by overlays. Colors, photographs, continuous tone illustrations, type, and line work can be mechanically keyed and separated for the printer. Methods of preparing mechanicals vary between industries and sections of the country.

Orthographic Projection. This method of viewing an object in space is the foundation of technical drawing. In multiview and axonometric drawing, it presents relationships as they really exist, not as they appear to exist. There are two distinct methods of orthographic projection:

1. The object is projected onto a flat surface. Several of these surfaces are then laid end to end, much like an unfolded cereal box.
2. The object is created in three dimensions in a particular orientation. All of the object's three-dimensionality exists in any of the orthogonal views.

These views are then conveniently arranged on the drawing surface.

To use an illustration in its most powerful manner you must constantly be aware that the object *is* three-dimensional, no matter what the particular viewing orientation is. To render it as a flat diagram is to cheat yourself of most of the object.

Papers (ES). Papers can be clear as glass or stiff as plywood. They can be organized by:

1. The weight, configuration (grain), strength, and opacity of the body of the paper.
2. The color, texture, tooth, grain direction, and sensitivity of the surface to various media.

When a sensitive surface is desired, but with sufficient body to be able to stand abuse, laminates called *boards* are made. The types of papers available for the various tasks involved in production of technical illustration are:

When designing:
flimsy tracing paper
ledger paper
vellum
kraft paper
anything available

When illustrating:
vellum
prepared cloth
prepared paper
drawing film
illustration board
electronic "paper"

Parallel Rule (ES). The parallel rule, which is laid across the entire drawing board, can either be a blessing or a hindrance in the production of illustration. It is more accurate than a T-square and quite a bit less expensive than a drafting machine. The parallel rule can change how you draw. It can also be clumsy, picking up anything in its way. And because it traverses the entire drawing surface, it is difficult to use part of the table as a reference area. Since the rule is always horizontal, an adjustable

triangle is normally used to form angles.

Paste-up. An illustration that is ready for the camera has often been "pasted" together from separate pieces of art, type, and photographs. Originally, these items *were* pasted, but today methods of assembling the paste-up include:

1. Wax—if the paste-up has to be changed, if a less than permanent bond is required, or if positioning is needed.
2. Rubber cement—if a semipermanent bond is required and if positioning is needed.
3. Spray adhesive—if a permanent bond is required and positioning is not needed.
4. Tapes—black masking, white masking, and plastic mending tapes for quick mounting of large pieces.

The efficient technical illustrator does not waste time producing art totally free of correction. If mistakes are made, they can be corrected with paste-up techniques. Not only can the camera be fooled, but the negative can be worked to make a final correction.

Pens (ES). A drawing done in ink is a durable document. Pens for ink work have evolved into reliable and accurate instruments. Pens may be classified into three categories (Figure 2-26):

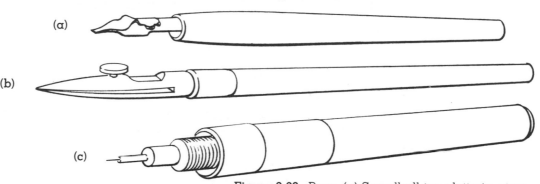

Figure 2-26. Pens. (a) Speedball-type lettering pen. (b) Ruling pen. (C) Technical pen.

1. Nonreservoir pens with points for lettering and sketching.
2. Reservoir pens with changeable points for sketching, drawing, or lettering.
3. Reservoir pens whose points are hollow cylinders with an internal metering rod. These are called *technical pens*.

The technical pen has replaced the ruling pen for most ink work, as it produces consistent line weights without constant adjustment and refilling.

Pencils (ES). Wooden pencils can be obtained in the same lead designations as the solid graphite used in lead holders (see Table 2-1). Since lead holders require small-diameter stock, the benefit of wooden pencils lies in their specialty applications. Excluding those used for general drawing, they include:

1. Very soft, dark, and smooth graphite in the 6B–3B range.
2. Specialty shapes, such as flat leads for chisel lines.
3. Colored pencils, including wax-based, pastels, and nonreproducible blue.
4. Special marking pencils, such as china markers, grease pencils, and lithographer's pencils.

Perspective Projection. Because the drawings of technical illustrators must be accurate and comparable, each illustrator must understand the nature of mathematically correct perspective. The illustrator can draw in perspective in three ways:

1. Manually, using only a scale, a straight edge, and pins for vanishing points.
2. Using aids such as perspective grids, tables, and lineads.
3. Using a machine, by which the operator either traces the orthographic views or enters the object as a set of points.

Aids and machines can be of great help. They can relieve much of the tedium of setting up and laboriously plotting complex forms. However, these perspective aids are no substitute for personally knowing how to draw perspective. The more an illustrator knows about perspective, the greater use these aids can be.

Photo-type (ES). The illustrator needs to know how to keyline, specify, and copyfit output from photo-type machines and computer-aided typesetters. Photo-type is produced on photographic sheets or strips from timed exposure through a visual font. Electronic type is produced on photographic film or sensitized paper from a visual display of the type in electronic memory.

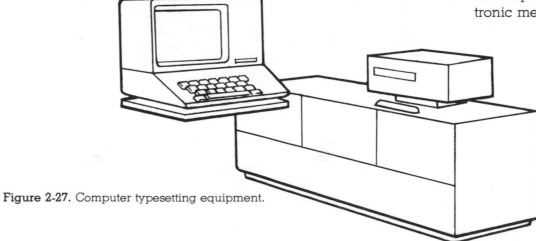

Figure 2-27. Computer typesetting equipment.

Figure 2-28. Splines.

Proportional Aids (EE). It is common for the illustrator to work from drawings at different scales. The easiest method for reducing or enlarging a drawing is to make a photo-enlargement or reduction. On those occasions when a photograph isn't practicable, the use of proportional aids makes scaling easier. *Proportional wheels* are like circular slide rules for figuring enlargement or reduction factors or percentages. *Proportional dividers* do the same thing by proportioning the divider's legs.

Ruling Pen (EE). Even though the ruling pen has been replaced by the technical pen for inking, it still has many applications. Paints and dyes that would not work well in technical pens do fine in ruling pens. Acrylic paint is a disaster in a technical pen but flows well in a ruling pen. Ruling pens function

best when they reflect the "set" of the illustrator's hand. Only when executing a very fine line, say, 4×0 or thinner, would an illustrator need to use a freshly sharpened ruling pen.

Splines (ES). Extruded acrylic rails or splines are held in place by lead weights (Figure 2-28) to aid in the drawing of smooth curves of considerable length. Some splines have lead inside, as a bendable core, but they are limited in the sharpness they can follow. The technical illustrator must be able to fit several pieces of a curve together in order to produce a smooth curve.

Swipe File (ES). A swipe file is a collection of drawings, photographs, figures, and so on that illustrators can use as a reference to fire their imagination. A swipe file must be organized or it will be useless. One useful method is a three-ring binder with document protectors for the examples, which can be

organized by subject or technique: people, buildings, transportation, appliances, line drawings, wet media, marker illustration, and so on. You have to be resourceful to build your swipe file *without* cutting up books or magazines. Commercial "clip files" are also available on a variety of subjects.

Tapes (ES). Tapes used in illustration are of two kinds: tapes that secure, and tapes used for drawing. Tapes that secure are usually drafting, masking (tan, black, and white), or plastic tapes. Drawing tapes, often called pressure-sensitive tapes, are used in place of ink or pencil lines. Widths from 1/64" to 2" and lengths from 324' to 648' are common. These tapes are available in black, white, or individual hues in matte, glossy, or transparent material. Drawing tapes can and do save time over inking. When the surface

of the illustration is textured or damaged, drawing tape may be the only answer.

Transfer Films (ES). Many pressure-sensitive film products are available to aid the illustrator. These transfer films include:

1. Typographic sheets of full or partial fonts (depending on the type size), available in most type styles found in photo-type or hot type.
2. Overlay screens or textures that approximate photographic screens, shades, or line tints.
3. Overlay colors to approximate gels, filters, and Bourges sheets.
4. Specialty sheets containing logos, register marks, arrows, dots, trees, or template parts.

For the illustrator who is without the services of a well-equipped photo lab, or who is unwilling or unable to contract for such services, these pressure-sensitive products are the only way to produce cels, overlays, color keys, color separations, and transparencies.

T-square (EE). The T-square is still the basic tool of the technical illustrator, although many of its functions have been replaced by other mechanical aids. The T-square provides a rapid way to construct illustrations, check paste-ups, and do keylining. The illustrator should consider the following in the selection of a T-square:

1. The head should be as long as possible and should be attached to the blade by five fasteners with counterset nuts. The blade and head should be perpendicular to each other.
2. The blade should have plexiglas edges. Check to make sure that the edges are set firmly in the blade.
3. Sight down the end of the blade for straightness.

Speciality squares are available for specific applications. They include T-squares with:

1. An extruded aluminum blade with plexiglas edges.
2. A full plexiglas blade to allow the illustrator to view everything under the blade.
3. A stainless steel blade to use in cutting paper or board in cases where a plastic edge would cut or nick.

Triangles (EE). In combination with a T-square, triangles can be used for perpendicular and parallel construction. Professional grade triangles are die cut from extruded acrylic sheet. Lesser grades are router cut or are cast from a more brittle thermoplastic material. Triangles can be found:

1. Clear or tinted.
2. In sizes 4" to 18" as standard.
3. With angles of 45°–45°–90° or 30°–60°–90° as standard.
4. Adjustable from 45°–90° with built-in protractor.
5. With or without inking edges.

Generally, larger triangles are a better investment. However, miniature triangles can be used for inking in cases where larger triangles would be clumsy. Be careful of triangles that are a little thicker than normal, as inking pens will not work with them.

Vellum (ES). This translucent drawing paper is the mainstay of technical illustration. It is used for layout, visualization, overlays, and finished art. Quality vellum is 100% white rag of approximately 20 lb. weight. The surface is of plate finish, which takes ink very well but which loads rapidly with pencil. Some vellum is unprepared—that is, the natural resistive nature of the surface has been left unaltered. Others are prepared but still respond to a light rubbing of tracing powder, or *pounce.*

An illustration done on vellum is not permanent. Vellum will crack and tear in a flat file that is opened, searched, and closed on a daily basis. For this reason, many companies keep vellum drawings in a *historical file,* using microfilm or microfiche to make day-to-day prints.

Wax Coaters (EE or CE). A method of positioning art and type that is less permanent than spray adhesive or rubber cement is coating the underside of the piece with a thin layer of special wax. This wax works well to hold art in place for photographing or platemaking but soon dries and releases unless well burnished. The wax can be applied with a small hand roller or with a mechanical waxer that can handle an entire sheet at a time (Figure 2-29).

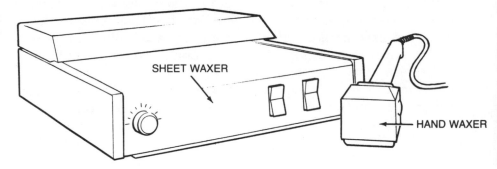

SHEET WAXER

HAND WAXER

Figure 2-29. Wax machine and hand waxer.

List of Terms

acetate	filing systems	minor axis
adhesives	fixatives	orthographic projection
airbrush	flat file	papers
axonometric projection	frisket	parallel rule
axonometric scales	graphic devices	paste-up
Ben Day process	halftone screen	pens
blueprinting	handrests	pencils
burnishers	hue	perspective projection
camera lucida	illustration board	photo-type
capital equipment	inks	pounce
cathode ray tube	interactive computer-	proportional aids
chroma	aided graphics	ruling pens
color systems	keylining	splines
color value	knives	stat camera
curves	lead holder	swipe file
drafting machine	lead pointers	tapes
drawing table	leads	technical pens
dry mounting	Leroy lettering	templates
dyes	lettering templates	thrust line
electric eraser	light table	transfer films
ellipse guides	major axis	T-square
erasers	markers	triangles
erasing shield	matte knife	trimming knife
expendable equipment	mechanicals	vellum
expendable supplies		vignette

Problems for Further Study

The problems in this section are designed to give you practice in using some of the aids and guides discussed in the chapter. The first problem asks you to translate orthographic shapes into axonometric views using your eye as a scale. The second problem asks you to make your own trimetric grid and use it to construct a view of a top manifold. The next problem presents axonometric construction ready for paste-up. The fourth problem calls for a photodrawing to be made from a photograph. The last problem requires a tracing to be made from a computer drawing underlay.

1. Here are several sets of detail drawings and axes for ellipse template practice. Use "eyeball" proportions to accurately capture the shape of each object.

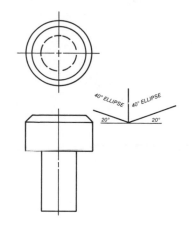

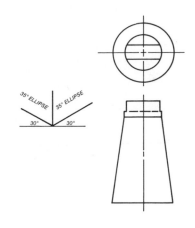

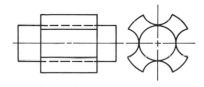

2. (a) Given the top and front views of a cube and a trimetric direction of sight, construct an auxiliary view of the cube, with foreshortened horizontal and vertical scales. Expand these scales into a ten-unit square grid. (b) Complete a trimetric drawing of a top manifold using the detail drawings provided and your grid.

(a)

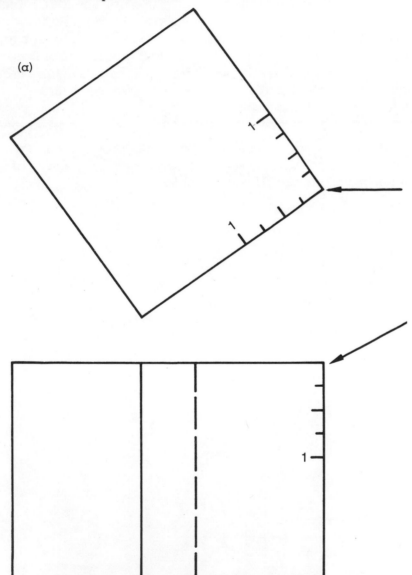

(b)

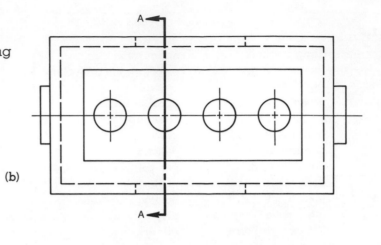

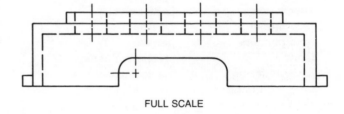

FULL SCALE

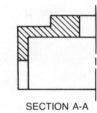

SECTION A-A

3. This is a sectional assembly drawing of a shaft and retainer and the rough construction of the parts. Assemble a paste-up in proper projection, making as many photocopies of the problem as necessary. Complete a finished illustration from the paste-up.

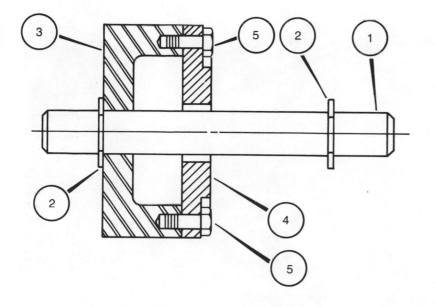

MK	RQD.	PART
1	1	SHAFT
2	2	EXTERNAL RING
3	1	HOUSING
4	1	COVER
5	6	HEX BOLT ½ – 11 × 1

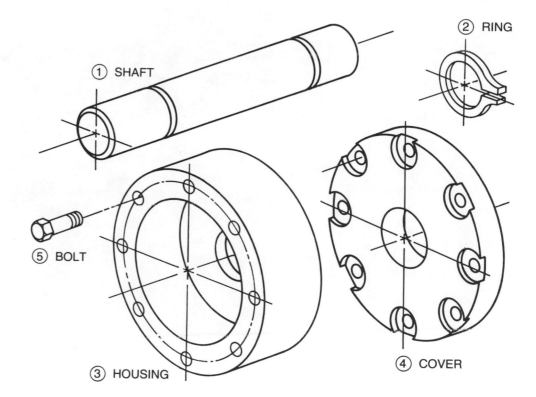

4. From this photograph of a hydraulic manifold and hoses, complete a photodrawing on acetate or vellum.

5. Here are the sectional assembly drawing and computer axonometric underlays for a collar assembly. Complete a paste-up of the underlay, and make a finished illustration on an overlay.

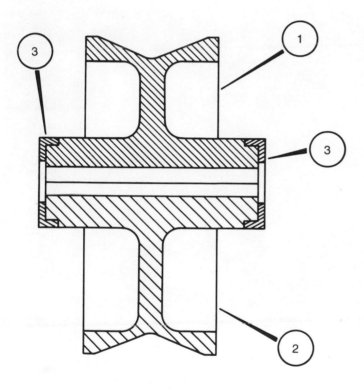

MK	RQD	PART
1	1	TOP CAP
2	1	BOTTOM CAP
3	2	COLLAR CLIP

COLLAR ASSEMBLY
SECTIONAL VIEW

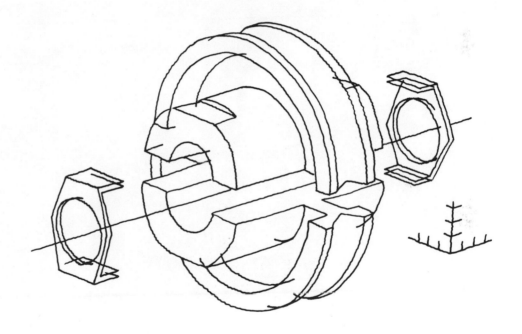

CHAPTER THREE Changing and Updating Illustration

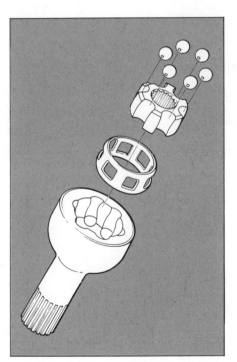

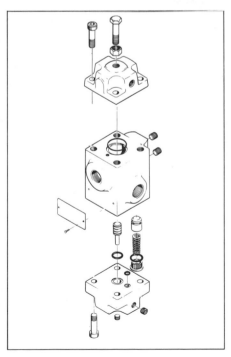

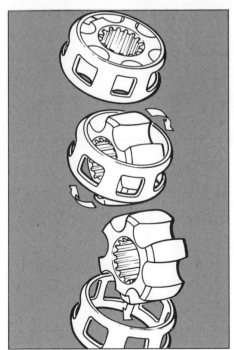

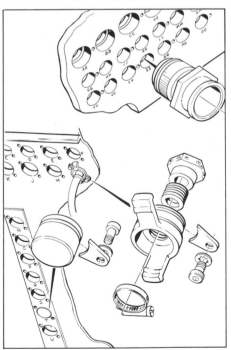

The first assignment that an illustrator may get is to change or update existing drawings. This type of assignment is a good way for a new illustrator to get experience with drawing practices and methods, although major revisions or actual "save jobs" often require the abilities of an experienced person.

One thing to keep in mind whenever reworking an illustration: Never redraw the illustration unless you absolutely have to. If the illustration was expensive to produce the first time, its *relative* cost can be reduced each time it is used in altered form without redrawing. Redrawing should be reserved for massive changes or corrections.

In making corrections, be careful of how many copies of copies that you make. It is easy to get three or four generations removed from the original and lose much of the detail in the drawing. This is especially true when making photocopies. If the original must remain intact, try to change a copy made directly from that original.

Keeping Track of Changes

Many companies have procedures for keeping track of the changes performed on each illustration. These procedures may include:

1. Requesting the change.
2. Verifying the change.
3. Determining cost of the change.
4. Documenting the change.

These controls accomplish several things: they keep unnecessary changes from taking up illustration time, they assure adequate information for making the changes, they provide budget information, and they provide a record of the changes made to each drawing.

Why Are Illustrations Changed?

Illustrations are continually changed. Sometimes the reasons are valid, and sometimes they are not. But the changes must still be made. Reasons include:

1. A change in design, assembly, or service. In this case, drawings need to accurately reflect the most current design (Figure 3-1). However, *every* design change need not require a change in the illustration. Unless the appearance or position of a part is altered, changing the part number in the bill of material can usually suffice.

2. The need to design an alternate or opposite-side assembly. This often requires only turning over or flopping the negative and making a reverse print. Slight changes can be made by covering or masking the negative before printing.

3. A change of purpose. An illustration for marketing might be altered to show a maintenance procedure. Or a maintenance illustration may be "cleaned up" or simplified to serve as a marketing illustration.

4. The need to redraw a poorly done illustration. Sometimes the only drawing you have of an object is an inferior copy, both in technique and reproduction. This can happen when companies go out of business or when long-abandoned designs are brought back into production. Some of these illustrations may have to be completely redrawn (Figure 3-2).

5. The need to correct a mistake. Mistakes do happen. They can be caused by carelessness, misinformation, a lack of information, or simply a lack of skill on the part of the illustrator.

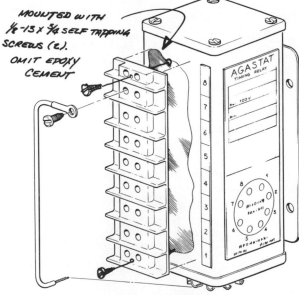

Figure 3-1. Changing an illustration by making a mask. In this case the epoxy must be removed and two screws added.

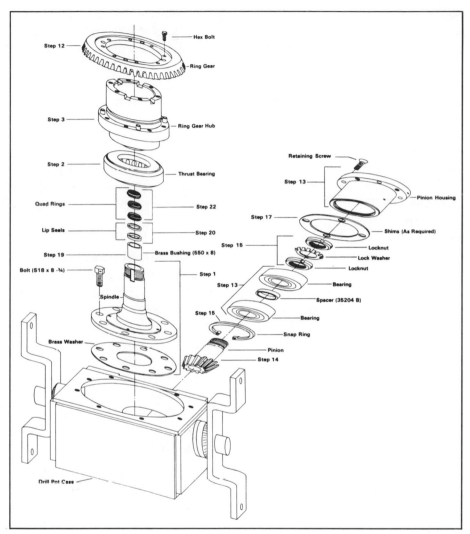

Figure 3-2. Changing a poorly drawn illustration. The ellipse exposure and tilt in this illustration are wrong. A major redrawing is required.

Materials Used in Changing Illustrations

Here is a list of some of the materials that can help you in changing and updating illustration:

transparent mending tape
white masking tape
bleed-proof white-out
secretaries' white-out
adhesive wax
rubber cement
spray adhesive
photo eradicator
roller
burnisher
photocopy machine
PMT prints
photoprints
blueprints
standard parts transfers
paper for masks

Methods of Changing Illustration

Illustrators use a variety of techniques, often in combina-

tion, when changing drawings. Normally they will pick the fastest, cheapest, yet most effective means to do the job. In general, changes are made in seven ways.

CUT-AND-PASTE ILLUSTRATION

Illustrations can be combined to make new drawings if parts are similar, if the angle of view is compatible, and if the techniques match. Scale can be adjusted by photographic enlargement or reduction. Keep in mind the angle of the parts and match ellipses for exposure and tilt. Once the axes in Figure 3-3 were drawn, parts were pasted along them. These parts were cut from several similar assemblies.

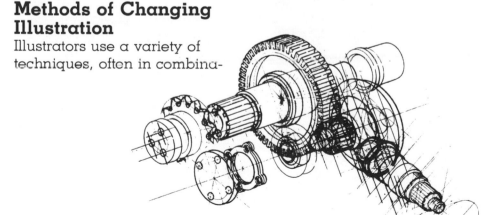

Figure 3-3. Changing an illustration by cut and paste. New or changed parts can be introduced along the drawings axes to create a new assembly.

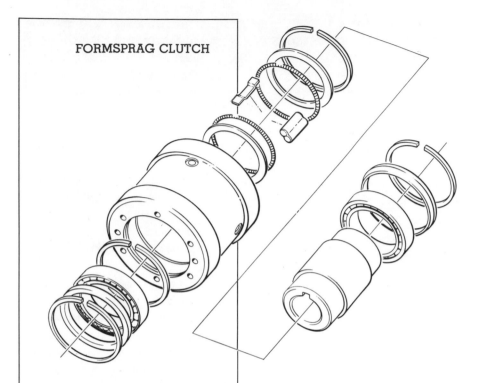

FORMSPRAG CLUTCH

IF THE NUMBER OF PARTS MAKE A *STRING* THAT IS TOO LONG FOR ALL THE PARTS TO BE PLACED AT A COMFORTABLE SCALE ON THE PAGE, *FLOW LINES* MAY BE OFFSET FOR BETTER BALANCE. RATHER THAN ALLOWING EACH PART TO FLOAT WITH AIR ALL AROUND IT, PARTS MAY BE STACKED CLOSELY TOGETHER. LINES TO THE REAR SHOULD NOT BE ALLOWED TO TOUCH LINES TO THE FRONT.

MASKING

If only a portion of the illustration needs to be omitted, a mask can be made to cover the appropriate parts. This method is better than cutting up the original, a practice that should be avoided. By using a mask, you get two illustrations—just flip the mask.

BLEACHING

Linework on photographic prints can be eradicated by the use of commercial liquid erasers. With this method, very small corrections can be made while leaving the surface workable for inking.

WINDOWING

If the finished illustration is to be copied photographically, cut marks and overlay edges can be ignored by the camera. Occasionally, however, a drawing needs to be corrected as an original piece of technical art. If the illustration is on vellum or film, the technique of windowing is particularly effective. A window is cut to include the af-

fected section, with matching material placed directly under the area to be cut. This way a *plug* is made that is the same size as the opening. The illustration is turned over and, with the aid of transparent mending tape, the plug is carefully positioned and secured. Both the rear and the cut marks on the front are then burnished. If done carefully, the plug cannot be detected.

ELECTRONIC REDRAWING

If the illustration was done on a computer graphics device, and if the drawing can easily be recalled, then portions can rapidly be deleted, moved, and repositioned. New parts can be drawn or assembled from existing templates and positioned to make new drawings.

ERASE AND CORRECT

Don't overlook the most obvious and often cheapest method of changing a drawing: erasing the error or part to be changed and redrawing it.

MANUAL REDRAWING

When all else fails, a new drawing must be produced. This method involves the most expense, in terms of both time and money. Keep as much of the original as you can to make a base print, and use an overlay to construct the new drawing (Figure 3-4). Look at the old illustration to see whether there is *any* information you can use to make your job easier, even if it is only distances between hole centers or relative ellipse sizes. You may also find that you can trace over parts of the old drawing.

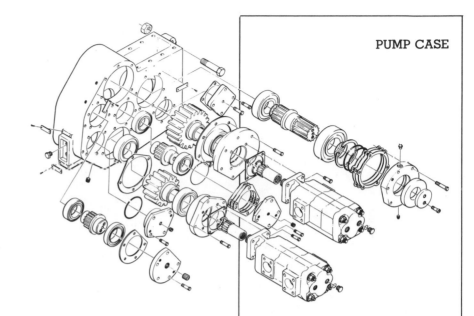

PUMP CASE

List of Terms

bleaching
cut-and-paste
design change
flopped negative
generations
masking
opposite-side assembly
redrawing
windowing

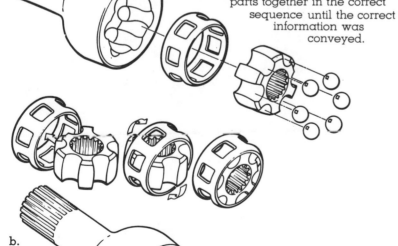

Figure 3-4. (a) This joint assembly would appear to accurately describe how the individual parts fit together. But in actuality the parts would not and could not be assembled as shown. (b) The correct assembly of the parts. Rather than a new drawing being made, the existing drawing was used as an underlay, fitting the individual parts together in the correct sequence until the correct information was conveyed.

THE CHOICE OF AXONOMETRIC VIEW OR ORIENTATION PLAYS AN IMPORTANT ROLE IN WHAT CAN BE SHOWN EFFECTIVELY IN THE DRAWING. IF, AS IN THIS CASE, A SHALLOW INCLINATION IS CHOSEN FOR ONE FACE, THE PARTS CAN BE STACKED CLOSELY TOGETHER, MAKING FOR A MORE COMPACT ILLUSTRATION. YOU CAN CHOOSE NOT TO EXPLODE CERTAIN PARTS IF THEY WOULD MAKE THE DRAWING TOO LARGE. THE TWO HYDRAULIC PUMPS AT THE LOWER RIGHT CAN BE EXPLODED IN ANOTHER ILLUSTRATION.

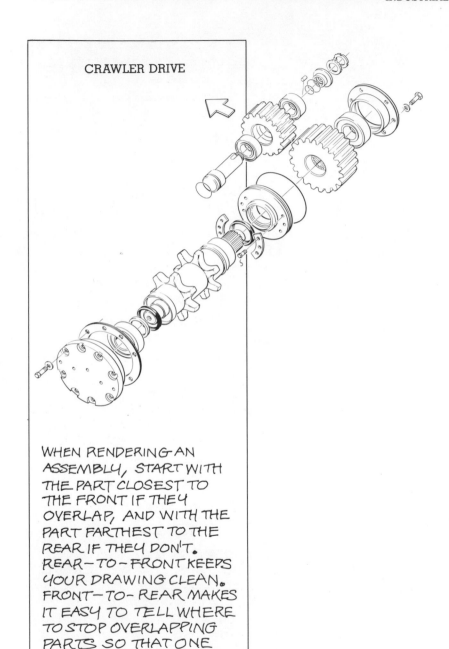

CRAWLER DRIVE

WHEN RENDERING AN
ASSEMBLY, START WITH
THE PART CLOSEST TO
THE FRONT IF THEY
OVERLAP, AND WITH THE
PART FARTHEST TO THE
REAR IF THEY DON'T.
REAR-TO-FRONT KEEPS
YOUR DRAWING CLEAN.
FRONT-TO-REAR MAKES
IT EASY TO TELL WHERE
TO STOP OVERLAPPING
PARTS SO THAT ONE
APPEARS IN FRONT OF
THE OTHER.

Problems for Further Study

Perform the indicated changes on the following examples, using the techniques discussed in this chapter. Make quality photocopies where necessary.

1. Make a mask to cover the indicated parts. Tape the mask so that it can hinge out of the way. Make copies of both the original and the altered illustration.

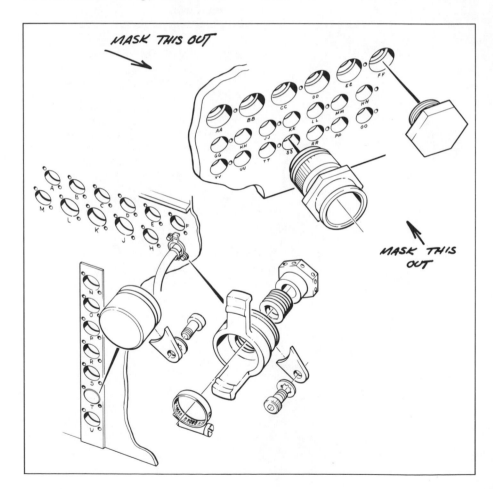

2. Remove the parts indicated and reposition the assembly to balance the parts. Make a photocopy of the corrected drawing.

3. Add the indicated parts to the assembly in the position marked. Make a photocopy of the new assembly.

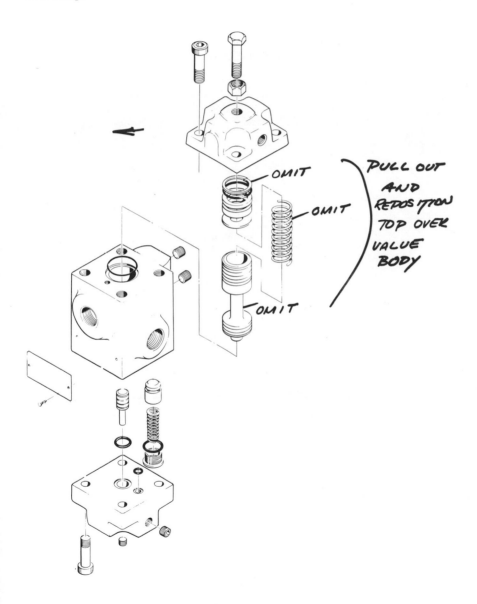

OMIT

OMIT

OMIT

PULL OUT AND REPOSITION TOP OVER VALVE BODY

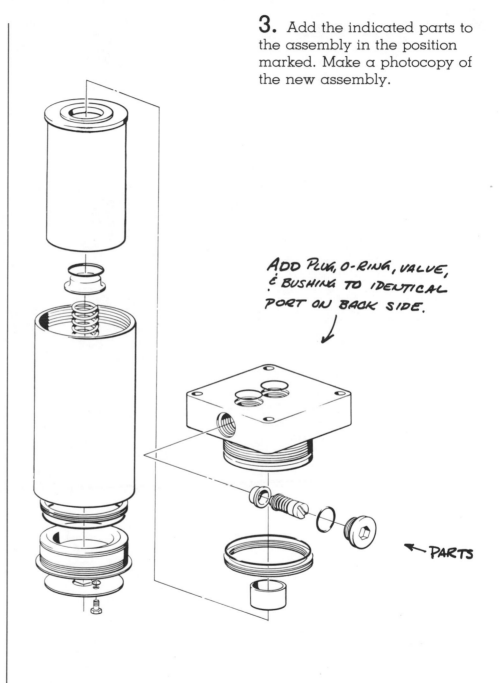

ADD PLUG, O-RING, VALVE, & BUSHING TO IDENTICAL PORT ON BACK SIDE.

PARTS

4. Make a copy of this problem. Using similar material, cut a window and plug to omit the part indicated. Tape and burnish the plug as described in the chapter.

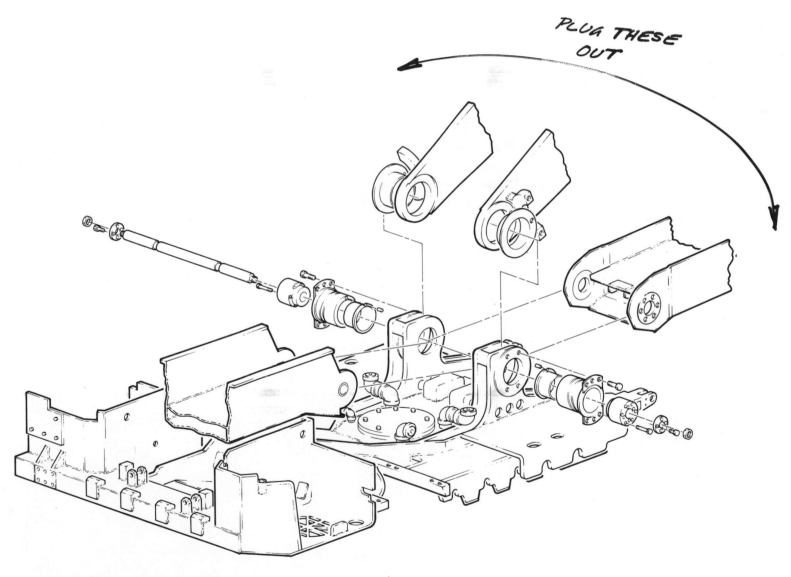

5. The only thing wrong with this illustration is the orientation of the ellipses. The correct orientation has been shown. Using any or all of the techniques discussed, alter the illustration to reflect the needed change.

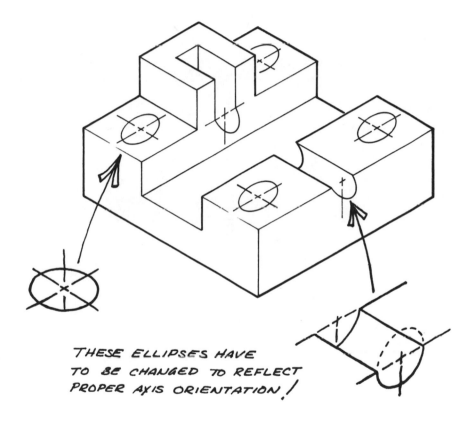

THESE ELLIPSES HAVE
TO BE CHANGED TO REFLECT
PROPER AXIS ORIENTATION!

CHAPTER FOUR Isometric Illustration

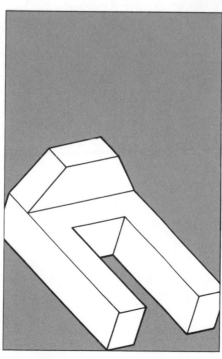

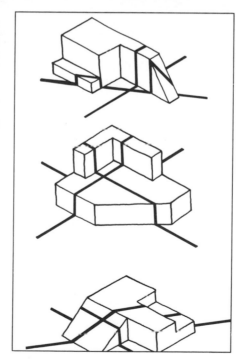

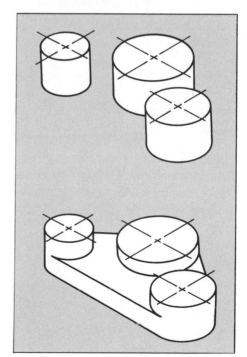

The best place to get started making your own industrial technical illustrations is with isometric drawing. Isometric is not the best method for all situations but it is the easiest in terms of construction. Its advantages include:

equal scale used on all axes
identical ellipses used on all
    isometric faces
common axis angle (30°)
large number of guides
    available

There are also sacrifices that may offset isometric's advantages:

least realistic view
inappropriate for long, narrow
    objects
can result in "isometric spread"

Isometric spread occurs because two lines remain parallel when your eyes tell you that the lines should be getting closer together. Lines that go off in the distance appear to get closer together in the natural world.

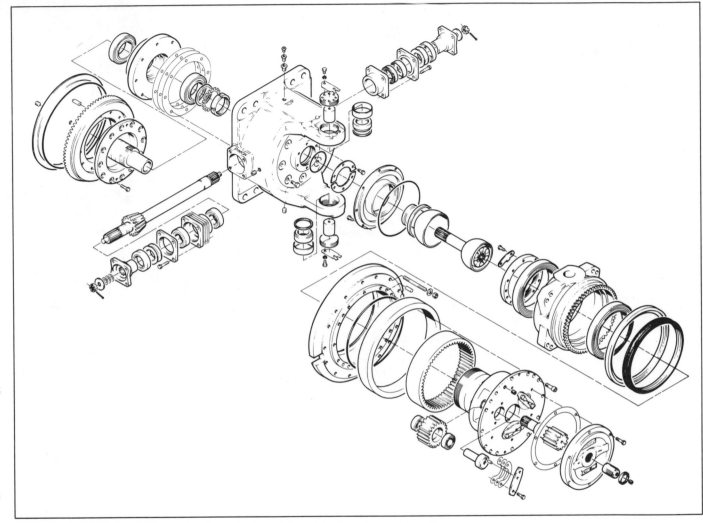

Figure 4–1. Technical illustration based on an isometric drawing. Notice that parts in front "float" away from parts in the rear by not allowing rearward lines to touch them. The same technique is used with the flow lines.

But since parallel pictorial drawing (axonometric drawing) is not natural—it lacks perspective—it can result in the lines appearing to spread or become farther apart as they recede. Dimetric and trimetric drawing are less inclined to suffer from this problem than isometric drawing.

For beginners in technical illustration, isometric drawing is the point of departure for more sophisticated pictorial drawing: dimetric, trimetric, and perspective. Most students learn isometric first, and indeed, a great deal of commercial technical illustration is isometric.

Figure 4-2. Isometric drawing. If the object were any longer, isometric spread would be more noticeable.

Figure 4-3. Which pictorial view? Take time to choose the most appropriate pictorial view. Consider the object's shape, prominent features, and your own drawing ability.

For these reasons it is the core around which skill in technical illustration is developed.

Isometric Drawing and Isometric Projection

The exact method of developing an isometric view is by *isometric projection*, a method discussed in detail in Chapter Five. Projection yields an image 82% the size of the original. Instead of making smaller scale drawings in this way, full-scale measurements are used. This results in an image approximately 1.25 times larger than an isometric projection, as shown in Figure 4-4. This larger drawing is called an *isometric drawing*.

In practice, isometric projections are not usually made, because of the need to scale all dimensions as well as the need to have two views of the object drawn at the same desired scale. Isometric drawings are made so that a convenient full scale can be used without having to project from scale views.

A 35° ellipse is used in isometric projection. An *isometric ellipse* is used in isometric drawing and is exposed slightly more than 35° (35° 16'). Most illustrators use isometric and 35° templates together in isometric drawing, increasing the number of available ellipses. Since the 35° ellipse is less exposed, it works well inside an isometric ellipse, reducing the difference

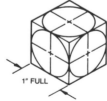

1" FULL   .82"

ISOMETRIC DRAWING   ISOMETRIC PROJECTION

Figure 4-4. Comparison of 1" cubes.

JON M. DUFF

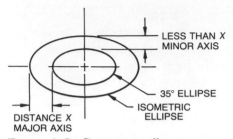

Figure 4–5. Concentric ellipses.

between inside and outside ellipses at the major and minor axes (Figure 4-5). Together, isometric and 35° guides can fit most applications.

ISOMETRIC DRAWING

Isometric drawing is characterized by equal full-scale measurement in the direction of the isometric axes *(base lines)*. Figure 4-6 shows the most common isometric view. All of the faces are equally exposed, making angles of 120° at the perpendicular edges of the

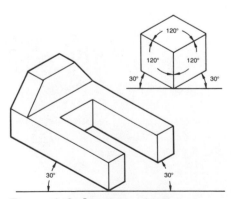

Figure 4–6. Isometric view.

planes. The *visual* angle that the bottom of an isometric box makes with a horizontal line is 30°. This allows you to use a 30° triangle or the 30° index position on your drafting machine to draw isometric lines to the left or right. Vertical lines remain vertical.

An *isometric line* is any line parallel to the vertical, right, or left base lines, as shown in Figure 4-7. You can measure directly along any isometric line. A *nonisometric line* is any line not parallel to any of the three base lines. These lines cannot be directly measured but can be constructed using several techniques discussed in this and the following chapter.

The key to isometric drawing is constructing lines on planes *in* the drawing, not simply drawing lines on the paper. The objects shown in Figure 4-8 represent planes in three-dimensional space. Two lines have been drawn across each object. Note that the lines don't simply cross the drawing; rather, they move up, down, and over the objects. In constructing an isometric drawing, measurements are moved up, down, and over the object just like the lines in Figure 4-8.

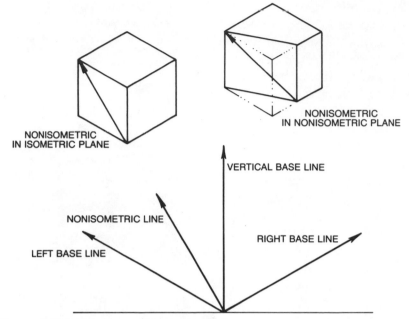

Figure 4–7. Isometric and nonisometric lines.

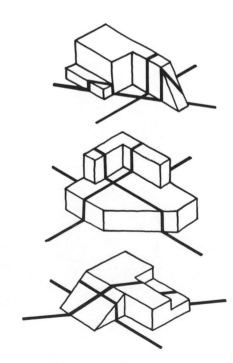

Figure 4–8. Drawing lines over three-dimensional objects. Don't just draw lines on the paper. These lines follow the planes of the object, further showing true shape.

AVAILABLE ISOMETRIC POSITIONS

Once you have chosen isometric drawing, there are still many positions available to orient the object. They include:

1. From above, as in Figure 4-9.
2. From below, as in Figure 4-10.
3. Along the long axis, as in Figure 4-11.

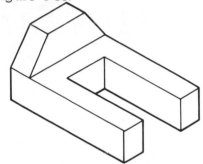

Figure 4-9. View from above.

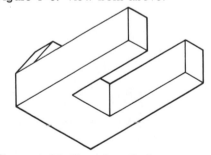

Figure 4-10. View from below.

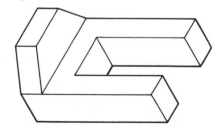

Figure 4-11. Long axis isometric.

In choosing a position, keep in mind the important features of the object. You usually don't want to hide important features. Still, there may be times that some features will be hidden no matter which position you choose. Then you may have to break out a portion of the object to show the obscured feature. Or you may want to make part of the object transparent in a phantom view.

Methods of Construction in Isometric

Methods used in isometric construction include boxing-in, offset, centerline, sectional, and grid.

THE BOXING-IN METHOD

The best way to start an isometric drawing is to build the form that encloses the whole object. In this way, you can see the limits of the object and more easily locate missing or difficult points. Use measurements taken from orthographic views and lay them out along the corresponding isometric lines to

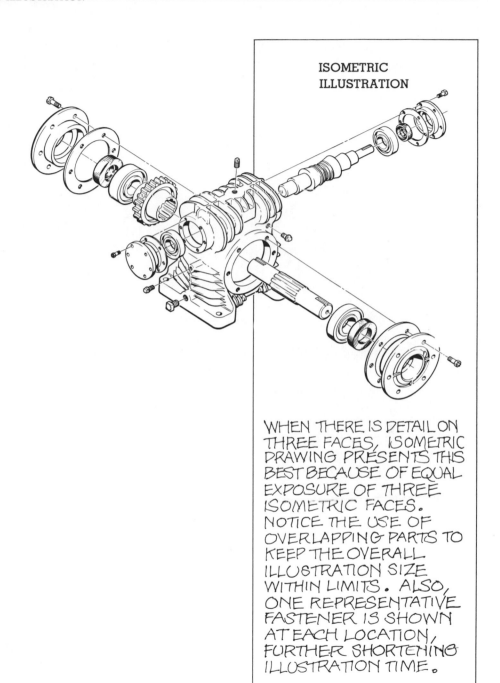

ISOMETRIC ILLUSTRATION

WHEN THERE IS DETAIL ON THREE FACES, ISOMETRIC DRAWING PRESENTS THIS BEST BECAUSE OF EQUAL EXPOSURE OF THREE ISOMETRIC FACES. NOTICE THE USE OF OVERLAPPING PARTS TO KEEP THE OVERALL ILLUSTRATION SIZE WITHIN LIMITS. ALSO, ONE REPRESENTATIVE FASTENER IS SHOWN AT EACH LOCATION, FURTHER SHORTENING ILLUSTRATION TIME.

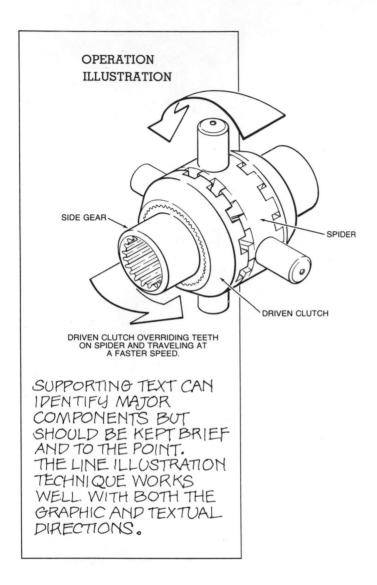

OPERATION ILLUSTRATION

SIDE GEAR

SPIDER

DRIVEN CLUTCH

DRIVEN CLUTCH OVERRIDING TEETH
ON SPIDER AND TRAVELING AT
A FASTER SPEED.

SUPPORTING TEXT CAN
IDENTIFY MAJOR
COMPONENTS BUT
SHOULD BE KEPT BRIEF
AND TO THE POINT.
THE LINE ILLUSTRATION
TECHNIQUE WORKS
WELL WITH BOTH THE
GRAPHIC AND TEXTUAL
DIRECTIONS。

build the overall form, as shown in Figure 4-12. Further define the details of the object within the box, measuring in from its sides. Darken the object lines when you are finished.

It is tempting to draw only a few lines and hope that you can draw the entire part. Yet if you keep your construction light and sharp, the many lines will aid you in your drawing. If your construction is fuzzy, smeared, or too dark, the lines may confuse you.

Difficult problems often call for the use of colored pencils. Use one color for centerlines, another for center points, and a third for cutting planes. Use colored pencils with hard lead, and avoid chalky or waxy pencils.

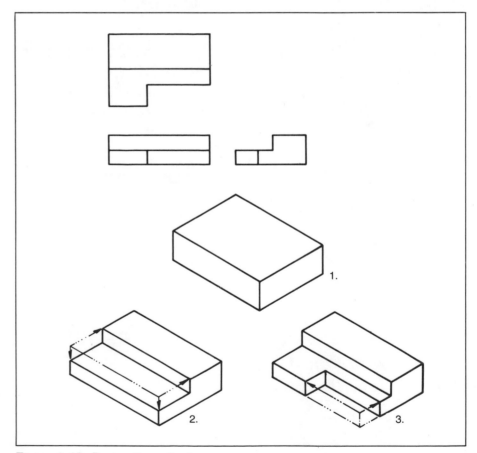

Figure 4–12. Boxing-in method.

Figure 4-13. Offset construction.

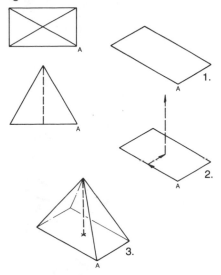

OFFSET CONSTRUCTION

Nonisometric lines can be drawn by using offset construction (Figure 4-13). The end point of the nonisometric line (A) is found by following it over, back, and up from known reference planes. When you are starting out, use the outside planes of the box that encloses the object as reference planes. As you become familiar with this type of construction, you can establish reference planes at any location, but always parallel to the isometric faces.

CENTERLINE CONSTRUCTION

Some objects do not lend themselves to boxing-in. Objects built around cylinders, cones, and spheres are better defined by their centerlines, as shown in Figure 4-14. When cylinders are connected by planes tangent to the cylinders, construct the cylinders first, then lay in tangent surfaces and their intersections.

SECTIONAL CONSTRUCTION

Compound and irregular curves are difficult to draw in isometric unless their sectional shapes are found first. A smooth outside line can be added to show the contour, as shown in Figure 4-15. If the cross-sectional shape is irregular, as opposed to the circular shape in the example, the shape is boxed-in and plotted as discussed in Chapter Five.

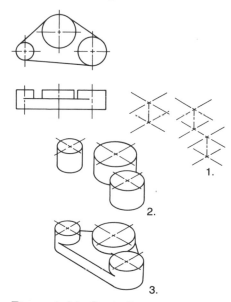

Figure 4-14. Centerline construction.

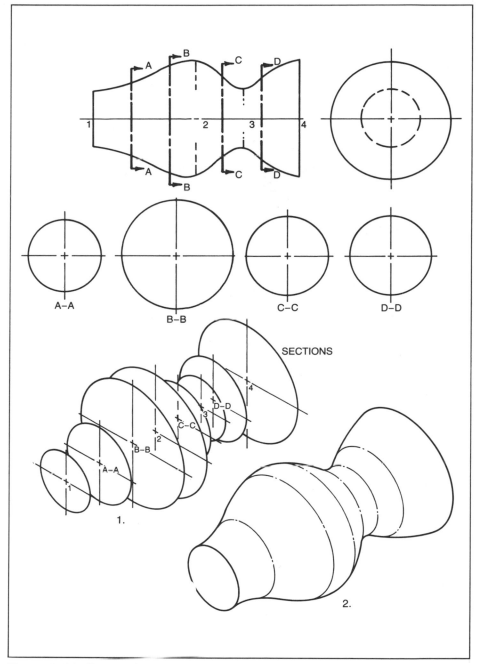

Figure 4-15. Sectional construction.

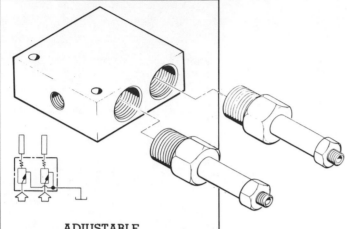

**ADJUSTABLE
RELIEF VALVES**

ONLY ONE OF THESE
VALVES WAS
CONSTRUCTED; A
PHOTOCOPY SERVED
AS THE BASIS FOR THE
SECOND. NOTICE HOW
THE BOTTOM OF THE
VALVE BLOCK "BREAKS"
AROUND THE FLOW
LINES, AS DO THE
INTERNAL THREADS.
THIS HELPS THE FLOW
LINE COME TO THE
FRONT AND THE OTHER
LINES MOVE TO THE
REAR.

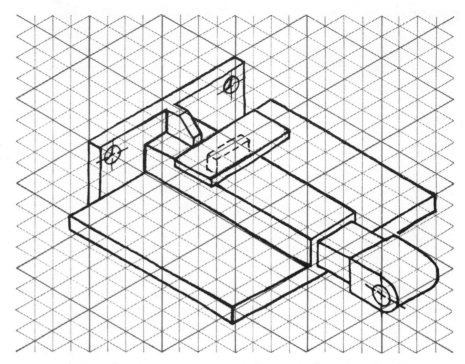

Figure 4–16. Freehand isometric sketch over isometric grid.

ISOMETRIC GRID PAPER

Isometric grid paper is not as helpful as dimetric or trimetric grids because of the ease of direct measurement in isometric drawing. Grid paper can be helpful in isometric sketching, however. Sketches such as that in Figure 4-16 should be done on an overlay sheet, keeping the grid clean for further sketching.

Shapes in Isometric

When you draw objects in isometric, you are practicing solid geometry. If you understand basic geometric shapes, how to draw them, and what happens to them when they are altered or cut, you can draw almost any object. Since this isn't a book on solid geometry—or descriptive geometry, for that matter—I have included only those shapes that you will run into most often. Study each of the shapes. Practice drawing them and cutting them in the study problems at the end of the chapter.

THE CYLINDER

A *right circular cylinder* has circular ends that are directly above each other. The height of the cylinder is shown by the dashed line in 4-17(a). Notice that the outside of the cylinder just touches or is tangent to the two ends. Horizontal cuts produce identical shapes down the cylinder (Figure 4-17b). Vertical cuts (Figure 4-17c) produce rectangular planes the height of the cylinder, with width determined by how far into the cylinder the cut is taken. Angled cuts (Figure 4-17d) are plotted by passing a number of vertical cutting planes through the orthographic view and laying off distances to the cut in isometric.

THE CONE

A *right circular cone* has a vertex directly above the center of its circular base (Figure 4-18a). Cuts made parallel to the base and perpendicular to the height reveal circles that remain tangent to the outside of the cone (Figure 4-18b). Cuts through the vertex (Figure 4-18c) produce triangular shapes. An angular cut, as in Figure 4-18(d), will make an ellipse. In the example, four points on the ellipse have been found by making use of lines on the surface of the cone (elements) that connect the vertex with the center points of the base. The distance across point 2 is *not* the minor axis. Point 2 is used to locate two points on the ellipse slightly above the minor axis.

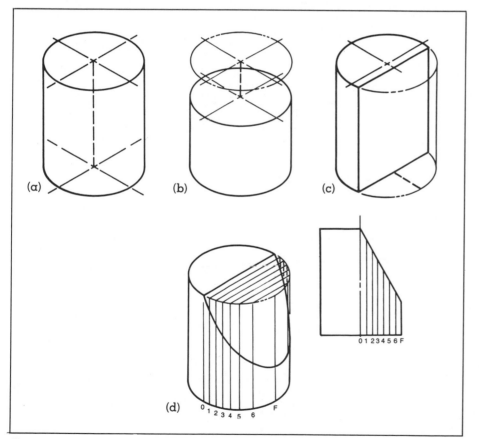

Figure 4–17. The cylinder.

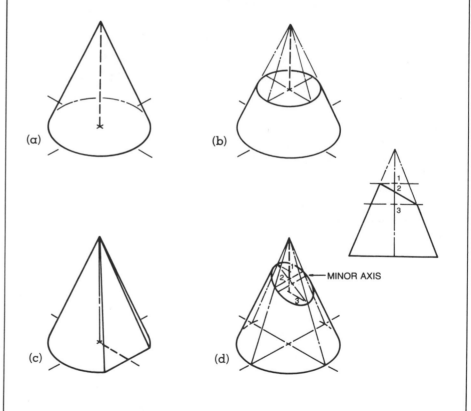

Figure 4–18. The cone.

HYDRAULIC VALVE
ASSEMBLY

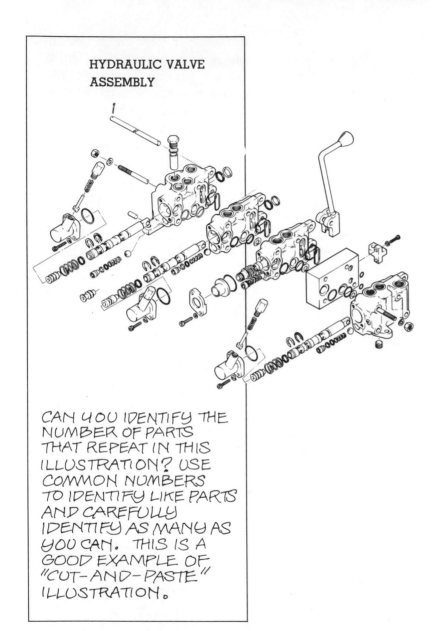

CAN YOU IDENTIFY THE
NUMBER OF PARTS
THAT REPEAT IN THIS
ILLUSTRATION? USE
COMMON NUMBERS
TO IDENTIFY LIKE PARTS
AND CAREFULLY
IDENTIFY AS MANY AS
YOU CAN. THIS IS A
GOOD EXAMPLE OF
"CUT-AND-PASTE"
ILLUSTRATION.

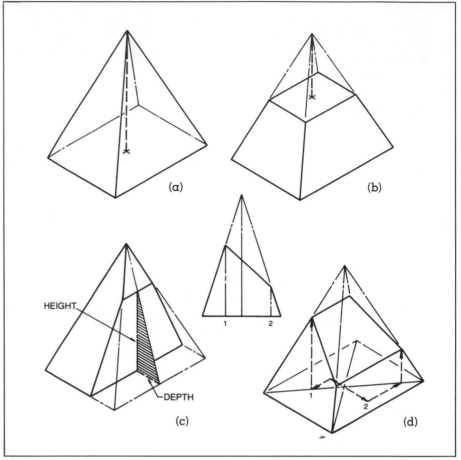

Figure 4–19. The pyramid.

THE PYRAMID

A *right rectangular pyramid* has a vertex directly above the center of the base (Figure 4-19a). Sections parallel to the base will reveal shapes similar to the base (Figure 4-19b). Vertical sections are found using the depth and height of the slice (Figure 4-19c). An angled cut is found by plotting the height where the plane cuts the corners of the pyramid (Figure 4-19d). Remember, you can only measure parallel to the isometric axes. This is a good example of using offset construction to find nonisometric lines.

THE SPHERE

To draw a 1″ sphere, construct horizontal and vertical 1″ ellipses around the same center point. The distance from the center along any of the isometric axes is assured to be ½″. Draw a circle with its center at the center of the ellipses and tangent to the ellipses at their tips (Figure 4-20a). The ellipses divide the sphere into eight equal parts and are used to draw a horizontal cut (Figure 4-20b). Vertical cuts (Figure 4-20c) also make use of these division lines. Vertical and horizontal cuts in isometric planes will be shown by isometric ellipses. Angled cuts (Figure 4-20d) are plotted from the orthographic view, using horizontal or vertical cutting planes to locate the major and minor axes.

THE TORUS

A *circular torus* may be thought of as many identical vertical circles all standing around in a circle. Thus, corresponding points on any two circular slices are the same distance from the center (Figure 4-21a). This fact is needed to plot shapes on the torus. A vertical slice through the center along the isometric axis will produce circular cross-sectional shapes (Figure 4-21b). Horizontal slices will produce circular rings (Figure 4-21c). Make use of the sections at the centerlines as well as the circular rings connecting top, center, bottom, and inside points when you do these sections. Angular cuts or vertical cuts at places other than the centers can produce some very odd shapes (Figure 4-21d). These shapes should be carefully plotted from the orthographic views of the torus.

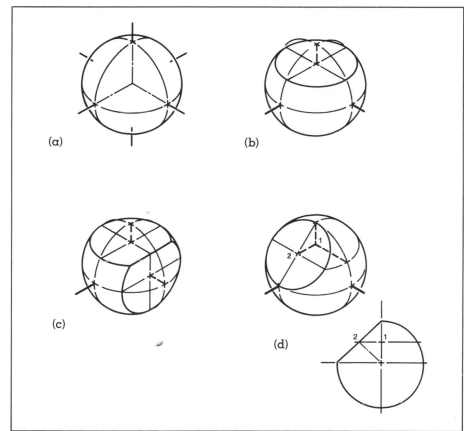

Figure 4-20. The sphere.

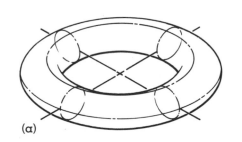

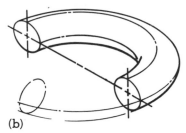

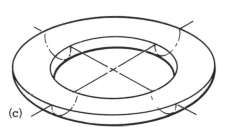

Figure 4-21. The torus.

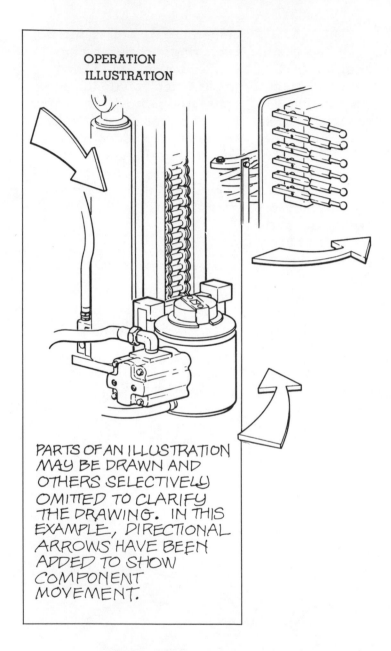

OPERATION
ILLUSTRATION

PARTS OF AN ILLUSTRATION
MAY BE DRAWN AND
OTHERS SELECTIVELY
OMITTED TO CLARIFY
THE DRAWING. IN THIS
EXAMPLE, DIRECTIONAL
ARROWS HAVE BEEN
ADDED TO SHOW
COMPONENT
MOVEMENT.

List of Terms

base lines	isometric grid paper	right circular cone
boxing-in method	isometric lines	right circular cylinder
centerline construction	isometric positions	right rectangular pyramid
isometric axes	isometric projection	sectional construction
isometric drawing	nonisometric lines	sphere
isometric ellipse	offset construction	torus

Problems for Further Study

The easiest way to learn isometric drawing is to *do* isometric drawing. Select some problems from this set that are appropriate to your level of ability. The first problems are the easiest, with difficulty increasing as you go along.

Do your construction in light, sharp lines, making use of colors where necessary, but always using complete and full linework. Once you have finished the construction, overlay a tracing sheet and darken your linework. Be prepared to turn in *both* the underlay and the overlay for evaluation. Refer to this chapter and to Chapter Five for construction tips.

1. Begin your practice in isometric drawing by producing finished line illustrations from the freehand sketches given in problems 1(a)–1(g). Use the tick marks to judge height, width, and depth measurements. Not only may you be required to make sketches from the actual object, but as an illustrator you may be given scale sketches by engineers or designers to "detail" into scale orthographic or pictorial drawings.

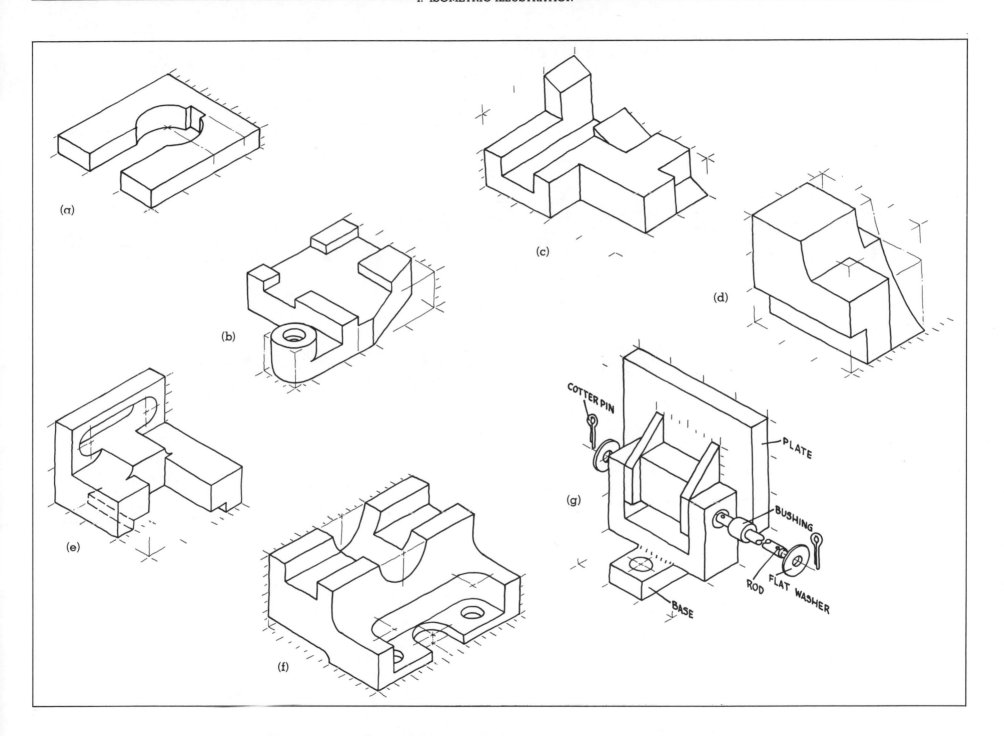

(α)

(b)

(c)

(d)

(e)

(f)

(g)

COTTER PIN

PLATE

BUSHING

ROD

FLAT WASHER

BASE

2–5. Produce isometric drawings of these basic shapes at full scale.

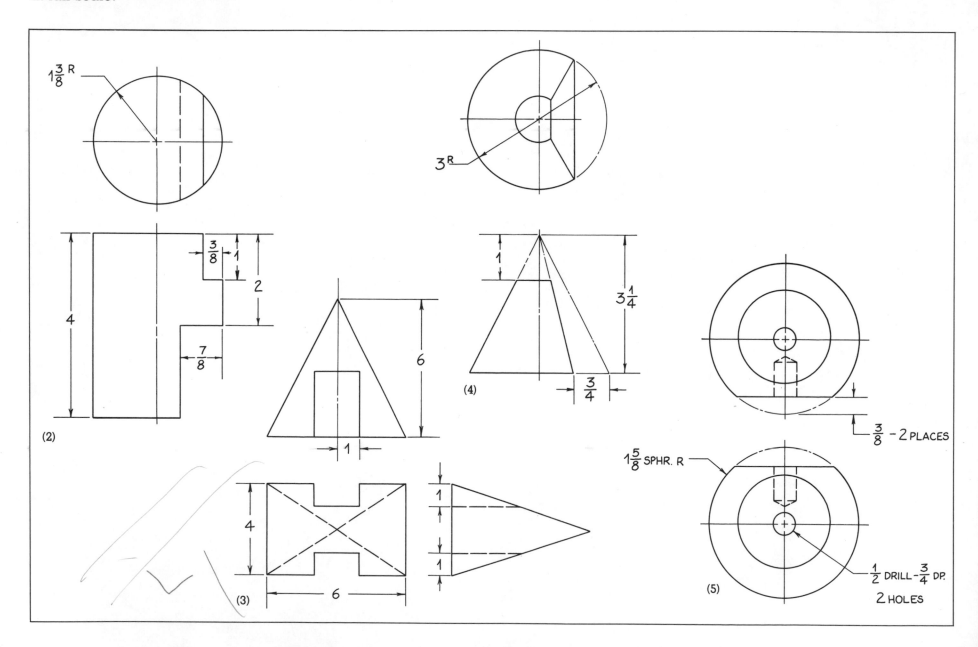

6. Set corner "A" at the corner of the isometric axes, making a half-scale illustration.

7. Use this scale drawing and your dividers to lay out a full-scale isometric drawing.

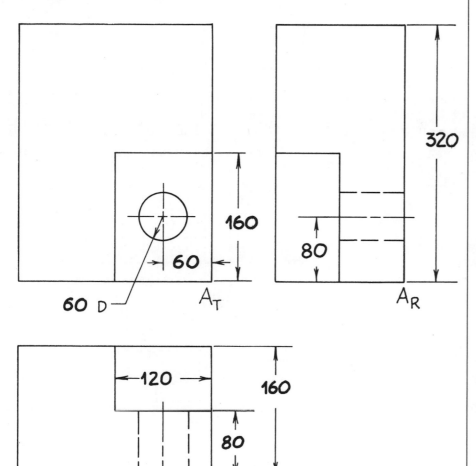

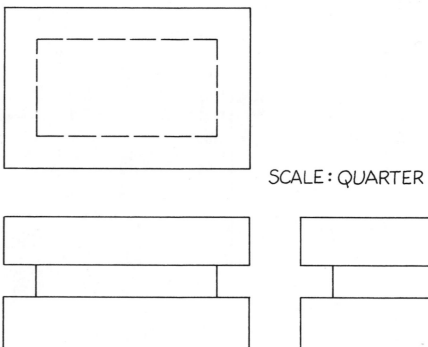

SCALE: QUARTER

8-20. Analyze each of these objects and determine the orientation that best shows its shape and features. Choose a scale that makes the drawing a manageable size, but not so small as to obscure detail.

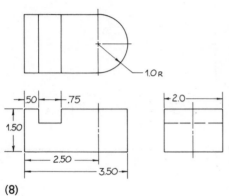

(8)

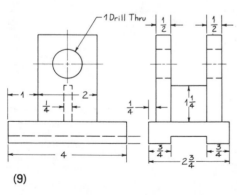

(9)

(10)

(11)

(12)

(13)

(14)

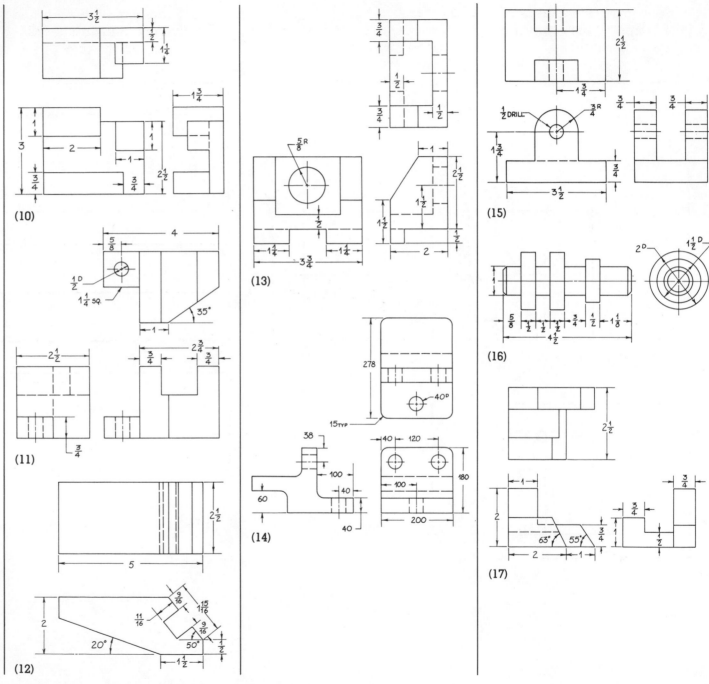

(15)

(16)

(17)

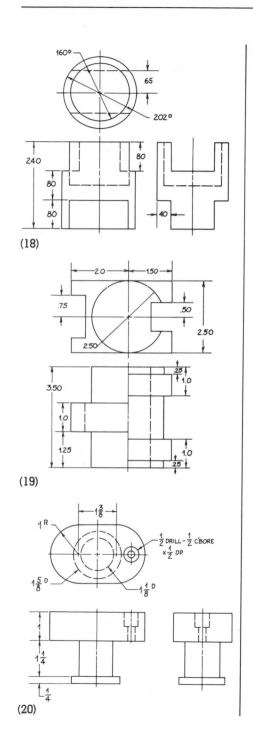

(18)

(19)

(20)

21. The clamping ring is drawn to scale. Construct an exploded isometric assembly drawing to include the ring, gasket, bolt, lockwasher, and nut. Execute the drawing using a twice-size scale.

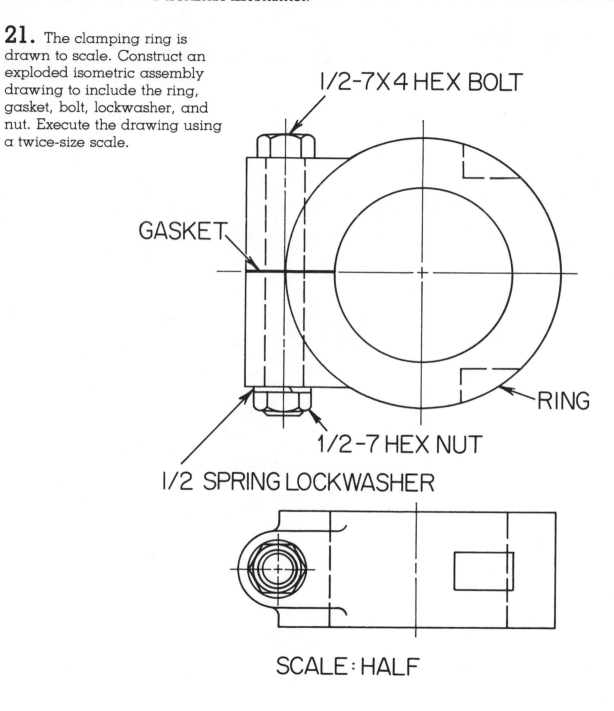

1/2-7X4 HEX BOLT

GASKET

RING

1/2-7 HEX NUT

1/2 SPRING LOCKWASHER

SCALE: HALF

22–24. Use these objects to practice drawing different isometric views of the same object. Produce at least two different views of each object.

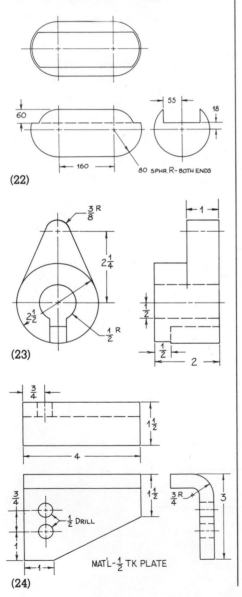

(22)

(23)

½ DRILL

MAT'L—½ TK PLATE

(24)

25. Make a four-times-larger drawing of the scale object shown. Note the sectional view, which shows the cross-sectional shape.

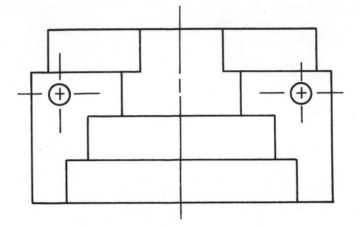

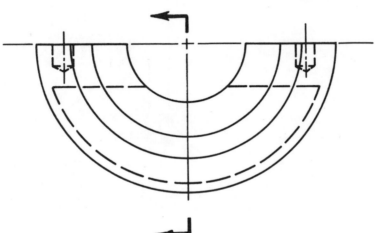

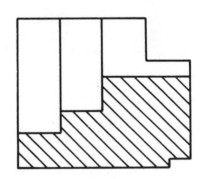

SCALE: 3/8" = 1"

26. Using the scale orthographic views, construct a full-scale isometric drawing. Orient the object so that you are looking down on the circular disk, putting the disk in the horizontal plane.

27. Complete a half-size isometric line illustration, turning the object so that the tangs (feet) are parallel to the left vertical isometric face.

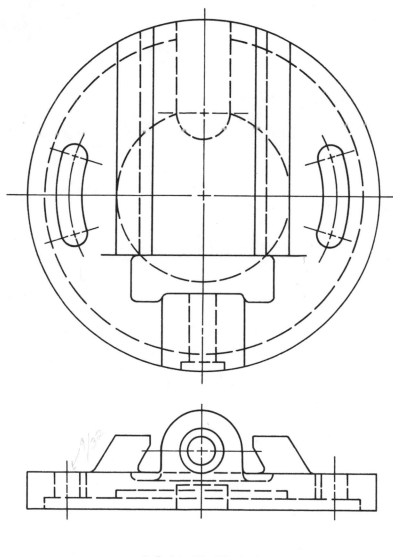

SCALE: FULL

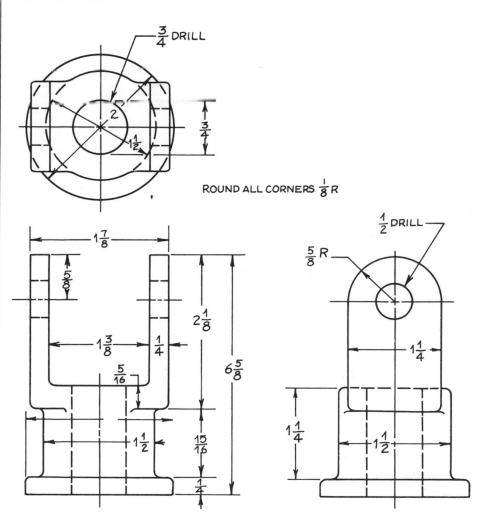

$\frac{3}{4}$ DRILL

ROUND ALL CORNERS $\frac{1}{8}$ R

CHAPTER FIVE

Advanced Axonometric Techniques

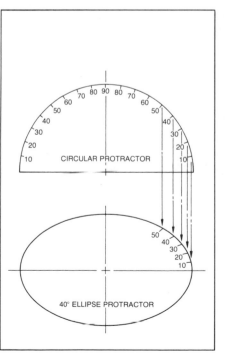

CIRCULAR PROTRACTOR

40° ELLIPSE PROTRACTOR

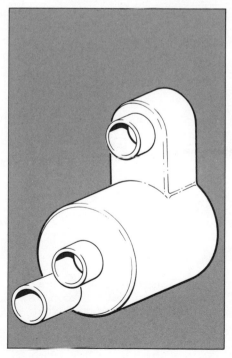

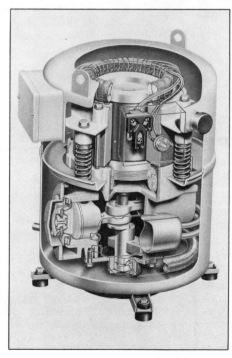

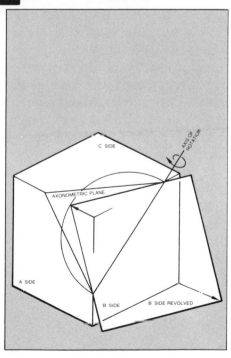

C SIDE

AXONOMETRIC PLANE

AXIS OF ROTATION

A SIDE

B SIDE

B SIDE REVOLVED

Technical pictorial drawing has been criticized for being too costly—not for lack of benefits received, but just for the high cost. Quality illustration *is* expensive. And the expense can be increased if the illustrator does not take care to be as efficient as possible. This efficiency must be built around a sound understanding of the capabilities of pictorial drawing and of the aids and guides described in Chapter Two that can be used to speed up the drawing process.

Most illustrators are trained to produce precise, exact, and technically accurate drawings. If a fastener is specified to have eleven threads per inch, the illustration will contain eleven threads per inch; if a gear is to have 62 teeth, the illustration will show all 62; if a spring has 15 turns, so will its illustration. Don't get me wrong! Every illustrator should be able to produce technically correct pictorials. The difference between producing affordable illustration and making do with something else (such as doctored engineering drawings) is knowing the *level* of precision

required to do the job. When is it necessary to show exact detail and when is it not? In order to deliver illustration at the lowest possible cost (and thereby promote future illustration assignments), the illustrator must include only that level of detail required by the particular illustration's use.

After you have mastered the basic axonometric drawing technique—isometric drawing—many factors become important. As a beginning illustrator, you may spend a great deal of time and energy just trying to figure the drawing out, leaving no time for deciding whether to show all the detail. The difference in level of detail can make a dramatic difference in the illustration's cost. Representative illustration can be half or quarter the cost of exact illustration. There is a place for both levels of detail. Each job must be analyzed in terms of the resources available. In general, the illustrator should think about:

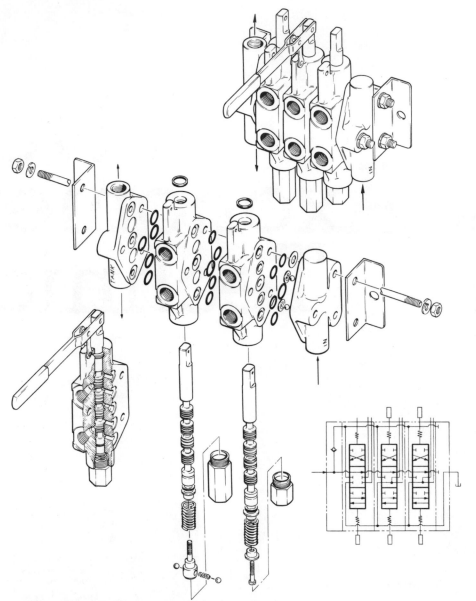

Figure 5–1. The range of technical detail that can be included in an illustration. This example shows an exploded pictorial, a sectional pictorial, an assembled pictorial, and a hydraulic schematic. An accomplished illustrator must often choose among these drawing examples.

Resources available:

1. Time.
2. Money.
3. Ability.
4. Information.

Intended use of the illustration:

1. Product development.
2. Marketing.
3. Parts identification.
4. Assembly/maintenance.

Reproduction (transmission) of the illustration:

1. Original art.
2. Photo-reduction.
3. Ozalid print.
4. Microfilm.
5. Electronic scanning.

Axonometric Theory

Axonometric drawing is a way of presenting the three orthographic dimensions of height, width, and depth on paper in one view. (In a principal orthographic view or orientation, two dimensions exist on the paper, with the third *in and out* of the drawing surface.) Although there are an infinite number of possible axonometric views, many have been standardized. This is largely due to the fact that commercial ellipse guides are available in 5° incremental steps, with a 90° ellipse as a cir-

cle and a 0° ellipse a line the length of the circle's diameter. For most drawing assignments, an illustrator need go no farther than a standard axonometric view.

But life is not always aligned with the axonometric faces. When an illustrator runs into angled or skewed lines and planes—ones that are inclined to the axes—he needs to know how to handle them. Knowing how axonometric views are formed is helpful in such cases.

Drawing vs. Projection

In an axonometric drawing, the illustrator measures directly along the axonometric axes, which results in an image that is slightly larger than that achieved by axonometric projection. Since most illustration is drawn as large as possible and then reduced photographically to a final size, the difference between projection and drawing is largely academic. It is the relationship of axis scale that is important, not the particular scale, since images can be enlarged or reduced quite easily.

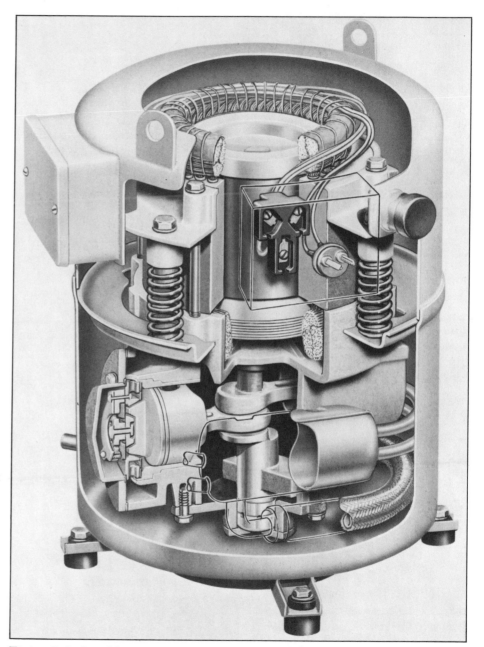

Figure 5–2. In addition to the drawing examples in Figure 5–1, there is the phantom technique, as used on this air conditioning unit. (Courtesy of Lennox Industries Inc.)

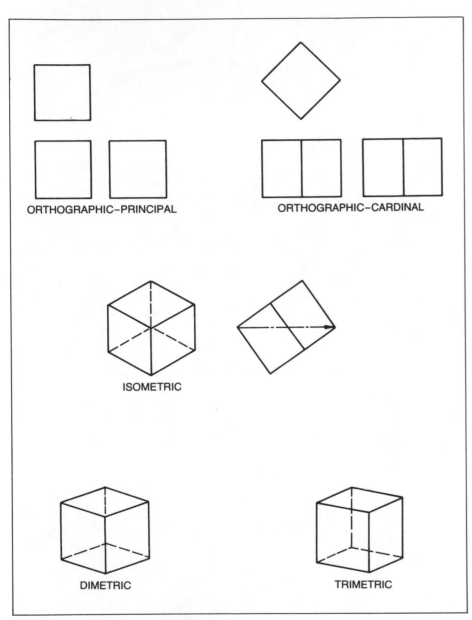

ORTHOGRAPHIC–PRINCIPAL

ORTHOGRAPHIC–CARDINAL

ISOMETRIC

DIMETRIC

TRIMETRIC

Figure 5–3. How views are achieved.

The Basic Axonometric Views

The position of the object in relation to the viewer determines how the object is inclined, as shown in Figure 5-3. When the object is squared-up, a *principal orthographic view* is obtained. If the object is rotated so that both planes seen in the front and side views are equally inclined, a *cardinal view* is achieved. When the object is tipped forward so that the body diagonal of the cube becomes a point in the front view, the view is *isometric*. Isometric pictorial drawing was covered in depth in Chapter Four. In an isometric view, the three scales are equally "ensmalled" or foreshortened. Keeping the two side planes equally exposed and rotating the body diagonal up or down causes the top plane to be more or less exposed than the sides. Such *dimetric views* are characterized by two equal scales. You might think of the isometric view as being the middle or central dimetric view, in which all three sides are equally exposed. Tipping the cube so that all three sides are exposed differently will yield a *trimetric view*. In a trimetric view, all three scales are unequal.

An isometric view, as we saw in Chapter Four, is easy to construct. No special scales are needed, and since two of the axes make visual angles of 30° with the horizontal, conventional triangles or index positions on a drafting machine can be used. However, isometric views often look the most contrived or forced. Of all the possible viewing angles available, only one is isometric. All the others offer varying views of the faces. For this reason many illustrators choose a dimetric view, especially when one of the faces of the object is very shallow (a 40°–40°–20° dimetric is a good example). Only two scales are needed and a more realistic view is achieved.

When the appearance of perspective is desired with the ease of direct axis measurement, a trimetric view is used. Each face has its own ellipse exposure; each axis has its own scale. Trimetric drawing is probably 10% more time consuming than dimetric drawing, and dimetric is about 10% more time consuming than isometric, when using direct construction. When projecting from given views, it takes roughly the same effort to draw in isometric, dimetric, or trimetric.

USING AN AXONOMETRIC DIAGRAM

The traditional method of making an axonometric illustration follows projection theory. Just as a principal view is created by projecting points onto one of the principal planes, an axonometric view is created by projecting the object onto an oblique plane, again in a perpendicular fashion.

Figure 5-4 shows the position of this axonometric plane. In order to use flat projection theory, the principal planes of projection must be rotated about the intersection of the principal and axonometric planes until all of the planes are co-planar. Points from the principal views are projected across this axis, with the axonometric view of the point located by the three intersecting projectors.

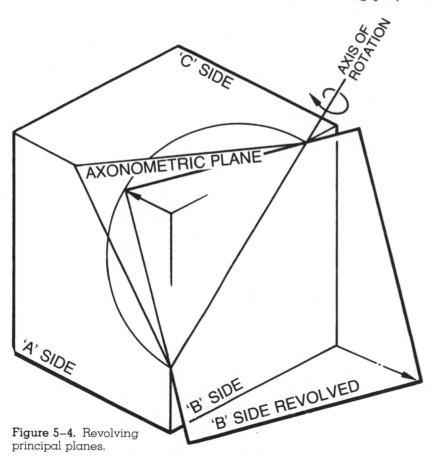

Figure 5-4. Revolving principal planes.

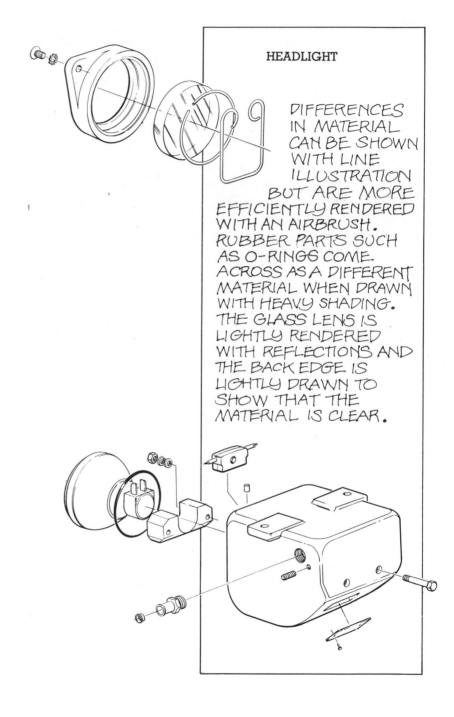

HEADLIGHT

DIFFERENCES IN MATERIAL CAN BE SHOWN WITH LINE ILLUSTRATION BUT ARE MORE EFFICIENTLY RENDERED WITH AN AIRBRUSH. RUBBER PARTS SUCH AS O-RINGS COME ACROSS AS A DIFFERENT MATERIAL WHEN DRAWN WITH HEAVY SHADING. THE GLASS LENS IS LIGHTLY RENDERED WITH REFLECTIONS AND THE BACK EDGE IS LIGHTLY DRAWN TO SHOW THAT THE MATERIAL IS CLEAR.

This projection is done on paper by constructing an *axonometric diagram*, as shown in Figure 5-5. Axes A, O, and X were established to present the corner of an axonometric box in a desired position. AX is constructed perpendicular to axis O; XO perpendicular to axis A; AO perpendicular to axis X. The midpoints of AO, OX, and XA are found and used for the centers of the semicircles shown.

Constructing perpendicular chords in each semicircle results in the revolved corner of the principal plane, as shown in Figure 5-4. Scales can then be projected as shown from the true edge, AR, onto the axonometric edge, AC.

The limitations of this method are obvious. The illustrator must have not only scale ortho-

graphic views of the object but orthographic views at the *same* scale. The illustrator cannot easily predict what will be seen and hidden when the axes are drawn. This method has almost totally been replaced by axonometric construction. The scales and grids used to make axonometric illustration are based on the axonometric diagram.

USING AUXILIARY PROJECTION
An axonometric view can be produced from auxiliary projection. Normally, it takes two auxiliary views to produce an interesting axonometric view (one that shows the three planes). If the object is rotated, as in the top view shown in Figure 5-6, a new front view has to be constructed.

Because the object is inclined toward the viewer in the top and front views, an axonometric view can be seen in the direction of the arrow. This does not free the illustrator from having to draw the orthographic views

Figure 5–5. Axonometric diagram.

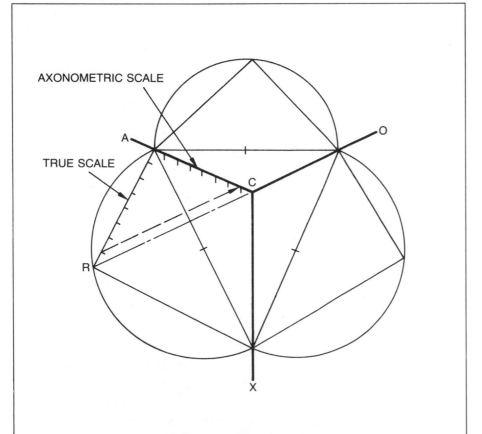

AXONOMETRIC SCALE

TRUE SCALE

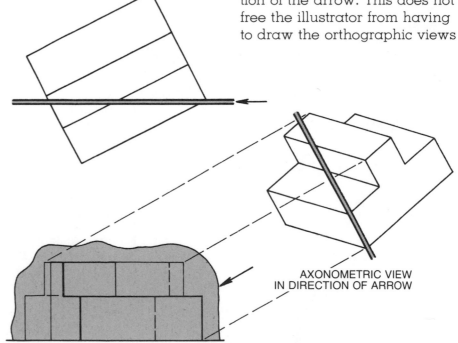

AXONOMETRIC VIEW
IN DIRECTION OF ARROW

Figure 5–6. Auxiliary projection.

Figure 5–7. Standard views.

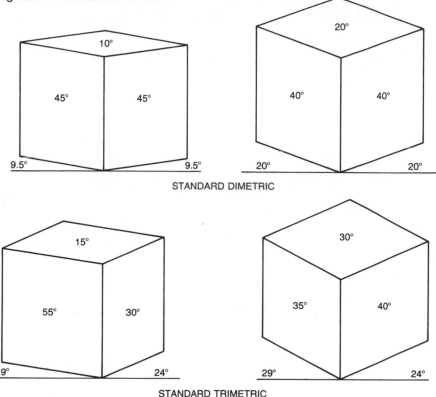

STANDARD DIMETRIC

STANDARD TRIMETRIC

at scale, however, This method is actually worse than the axonometric diagram because the views are not principal, as they would be on most engineering drawings.

From the auxiliary method it is simple to construct standard views, as shown in Figure 5-7. These views form the basis for grids. To make your own scales, look at Figure 5-8. A cube is drawn with the top and front views equally inclined to its sides. The true distance, which is laid off in the top view, becomes foreshortened in the front view. The scale is then ready for viewing from any angle. Notice that an isometric view is really a special case of all the possible dimetric views—the central dimetric view, as I noted earlier.

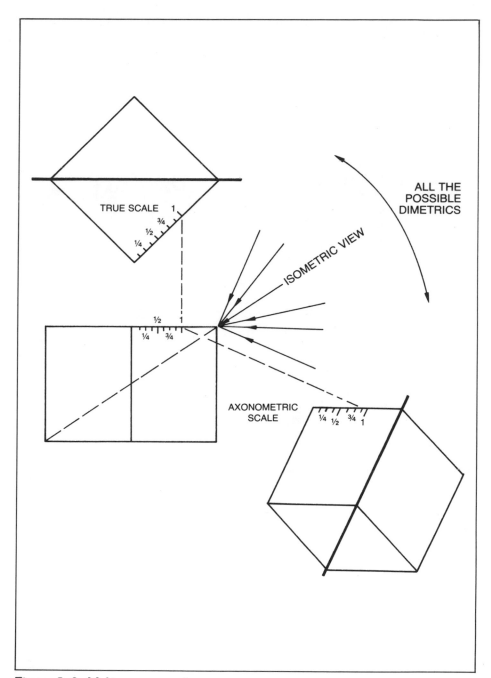

Figure 5–8. Making your scales.

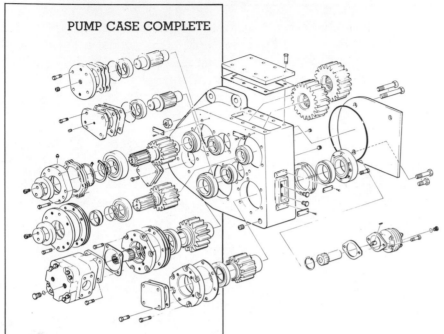

PUMP CASE COMPLETE

STRINGS OF PARTS MAY
NEED TO BE BROUGHT
OUT INITIALLY AT
DIFFERENT ANGLES TO
GET ALL OF THE PARTS
ON THE DRAWING. YOU
CAN AVOID CHANGING
THE DEGREE EXPOSURE
OF PARTS BY ROTATING
FLOW LINES AWAY FROM
THE AXES DIRECTIONS.
KEEP ALL PARTS
ALIGNED WITH THE
AXONOMETRIC AXES
WHENEVER POSSIBLE.
THIS KEEPS
CONSTRUCTION MORE
SIMPLE AND AIDS THE
EYE IN ASSEMBLING
THE PARTS.

Direct Construction from Grids or Scales

By far the fastest and most economical way of making axonometric illustration is by *direct construction*. In this method, the illustrator "builds" the object on the paper. To do so efficiently, the illustrator must think of the orthographic views as being three-dimensional. (Review the way the lines actually follow the planes as shown in Figure 4-8.) In direct construction, measurements are transferred from orthographic views onto axonometric axes. It is along these axes that true measurement can be made. Measurement in other directions requires special attention.

In direct construction, you begin by constructing the overall form that defines the object. Don't cheat yourself out of the full construction—it is easy to get in the middle of a problem, get lost, and not be able to figure out where you went wrong. Think of complete and accurate construction as your roadmap through the drawing. The investment of only a few minutes of drawing time can save hours of redrawing. For example, in Figure 5-9 a 1¼ × 2⅝ × 1⅝ block was first constructed using the vertical

and horizontal scales. (Scales may be of your own design or may be from one of the standard axonometric views.) Next, the top and support wall were defined within the larger volume. Finally, object lines were darkened to show the form. Hidden or dashed lines are usually omitted in axonometric drawing unless they are absolutely necessary to understand some aspect of the object.

STEPS IN DIRECT CONSTRUCTION

There are seven basic steps in direct construction.

First, *decide on the scale* that you will use. Most illustrators choose the largest practicable scale, because it produces the best illustration. If the drawing is too large, you will spend all of your time putting down graphite or ink. If it is too small, you may miss detail. A good method for determining scale is described later in this chapter.

Second, *decide on the view* that presents the object to its greatest advantage. Most pleasing axonometric views have a short or narrow face where ellipses are 15°–25°. Place features in this face that either have little detail (such as shaft ends) or have so much detail that it would take more time than is feasible to effectively present them.

Many companies decide on one or more axonometric views as standard for all of their technical illustration. Such standardization aids in drawing revision and compatibility. And, since several illustrators may work on different parts of a large assembly, their individual drawings need to be consistent so that a composite illustration may be made.

Third, study the information that you have on the subject. Whether it is in the form of engineering drawings, sketches, photos, or underlays, take some time to familiarize yourself with the information. Talk to yourself. Ask yourself questions about placement, shape, size, material, and so on.

Fourth, place your scales in a convenient spot on the drawing, or, if you make separate scales, carefully draft them out.

Fifth, construct the form of the object. Build the skeleton and then the overall form. Work from the large to the small; from the near to the far; from the general to the specific. Use a sharp, light pencil for construction (try a 4H) and keep it sharp!

Some construction may be done in nonreproducible blue pencil, although a drawing done completely in blue may be difficult to read. For this reason you may want to reserve blue and other colors for construction lines that have special meaning, such as bolt circles, centerlines, points of intersection, and so on.

Don't be afraid to write yourself little notes on the drawing, such as "bearing starts here" or "middle of gear teeth."

If there is considerable construction, you may want to try a series of overlays, each containing a portion of the total construction. One sheet of paper may become grooved, wrinkled, or soiled when it is on the board for a long time during a difficult construction. Using overlays can keep the "grub factor" down.

Sixth, check measurements as you go along rather than waiting until you've finished. Often subsequent measurements are chained, or built, on earlier ones. If the earlier ones are wrong, you may be wasting valuable time.

Seventh, render the form of the object. To do so effectively, you need to be aware of how the illustration is to be used, how quickly it must be done, and how it will be stored or transmitted. Choose the most effective rendering technique for the time, money, talent, and reproduction techniques available.

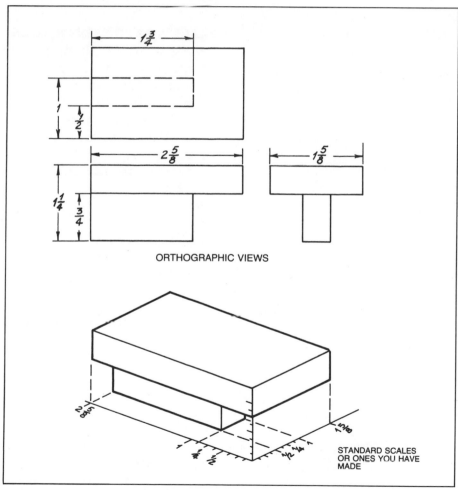

ORTHOGRAPHIC VIEWS

STANDARD SCALES OR ONES YOU HAVE MADE

Figure 5–9. Direct construction.

AN EXAMPLE OF DIRECT CONSTRUCTION

A full sectional view of a vacuum chamber is shown in Figure 5-10(a). It has three ports on the front and is made with a uniform wall thickness.

The axes chosen and their scale are used to first construct the overall form of the object, as shown in Figure 5-10(b). The appendage that includes port III is first constructed as if it were the same distance across as the large cylinder. The overall height, width, and thickness is laid out using the appropriate axis scale.

Next, the chamber is further defined, adding the form of the chamber's body and the centerlines and limits of the port cylinders (Figure 5-10c). Note that ports II and III lie on the centerline and are therefore easy to locate. Port I does not, so it must be found in a special way. The easiest method is the *offset method*, since the views in Figure 5-10 are drawn to scale. The distance that port I lies below port II on the centerline is measured, and then the distance from one end of the cylinder port to the other is measured.

Finally, the linework is completed to establish the solid form of the vacuum chamber (Figure 5-10d).

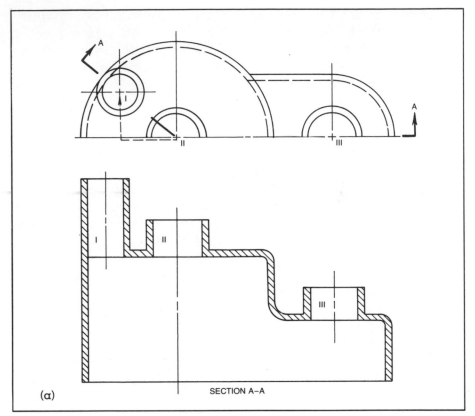

(a)

SECTION A–A

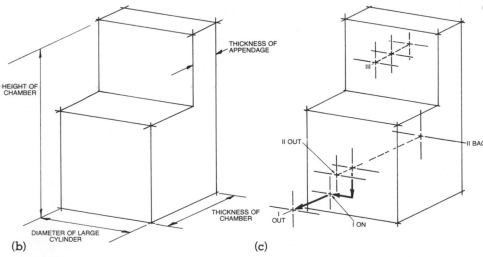

(b)

HEIGHT OF CHAMBER

DIAMETER OF LARGE CYLINDER

THICKNESS OF APPENDAGE

THICKNESS OF CHAMBER

(c)

II OUT

II BACK

OUT

I ON

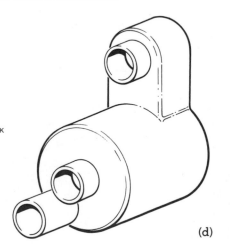

(d)

Special Construction Techniques

A number of special techniques are used in axonometric construction. Some are used each time an axonometric drawing is made. Others are used infrequently, but when they are used, they spell the difference between success and failure.

SELECTING THE RIGHT ELLIPSE

If you are working from a standard axonometric grid, the three face ellipses are usually marked on the sheet. If you are making your own grid, the easiest way to determine the correct ellipse for a given face is to first construct a square on that

Figure 5–10. Axonometric construction. (a) Sectional view of vacuum chamber. (b) Boxing in the form. (c) Locating centers. (d) Finished linework.

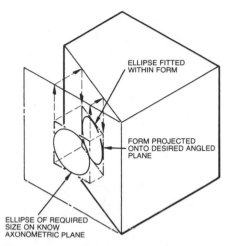

Figure 5-11. Selecting an ellipse.

face, then fit an ellipse within the square, tangent to the square on the centerlines. Note how this was done on an angled plane in Figure 5-11. Illustrators often have to approximate or use a "fudge factor" when the correct ellipse is not one of the standard ellipse exposures. Unless the desired ellipse is *very large*—say 6" diameter or larger—just use the closest one. A basic rule of thumb is to go *down* to the nearest standard ellipse if the ellipse is small. If it is large (2" and up), go *up* to the nearest standard ellipse.

Another way of determining the correct ellipse exposure is to use a commercial ellipse selector (or one you've made yourself). Figure 5-12 shows an ellipse selector for ellipses revolving in isometric planes. More often than not, an illustrator *approximates* an ellipse by knowing where the ellipse would fit between known ellipses. Figure 5-13 shows an approximation diagram for a 40°–40°–20° dimetric view. The illustrator knows that the horizontal ellipses will be 20° and that the ellipses in the right and left vertical faces will be 40°. Keeping this in mind, he can

find approximate ellipses by interpolation. Let's say an ellipse is desired at position A. How would the illustrator approximate that ellipse?

He would begin by noting that around the horizontal axis ellipse exposure goes from 40° to 75° and back to 40°. At position 1, the ellipse would be roughly halfway between 40° and 75°, or 55°. (The relationship is roughly linear except where it moves faster as the ellipse narrows.) Movement along arc 1-A is then from 55° to 20°. Position A is in the center, which means that the ellipse would be the largest of the two choices: 40°. This

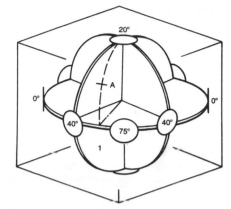

Figure 5–13. Ellipse approximation.

method frees the illustrator from having to calculate ellipse exposure, especially when doing designing or when no engineering drawings exist.

Of course, there are highly accurate methods of determining a more exact ellipse exposure. However, ellipse guides aren't that accurate. Principles of trigonometry can be applied to calculate exact values mathematically. Computer programs that produce axonometric images are based on these mathematics.

A method known as *circle projection* can also be used to determine ellipse exposure.

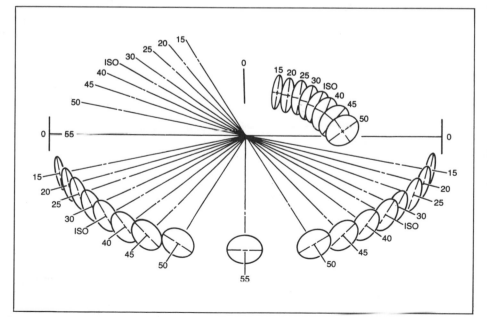

Figure 5–12. Isometric ellipse selector.

Circle projection is setting up a circular protractor for each angle measurement. To determine the proper ellipse for the nonaxonometric plane in Figure 5-14, a circle is drawn equal in diameter to the major axis of the ellipse. The plane is projected to the circle where the angle (20°) can be measured. A circle projection from this 20° ellipse can establish the ellipse exposure at point A (40°).

Better than circle projection, which has to be done each time, is a set of ellipse protractors to match your ellipse guides. They take only a few minutes each to make. Figure 5-15 shows how you can make an ellipse protractor by circle projection.

DETERMINING THE PROPER ELLIPSE FOR AN AXONOMETRIC FACE

If you are working on a view drawn earlier or if you know the way you want the object oriented, you can find the proper ellipse exposure graphically. Let's use Figure 5-16 as an example. Given dimetric axes converging at point O, what are the proper ellipse exposures for the faces?

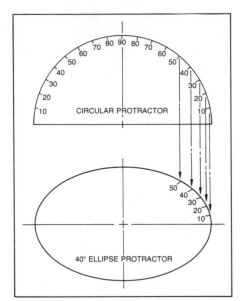

Figure 5–15. Protractor by circle projection.

To determine the ellipses, construct an axonometric diagram that revolves corner O on the three planes, as shown in Figure 5-4. To do so, construct arcs from the centers of 1-2, 2-3, and 1-3. On Figure 5-16 the arc has been shown in its most direct rotation for plane 1-2-0 and in the longer rotation for plane 2-3-0. Both rotations will yield a normal view of the principal plane.

Look down the end of the axis of rotation (lines 1,2 and 3,2) and construct the axonometric position of O and its revolved

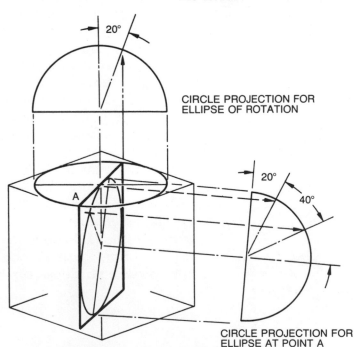

Figure 5–14. Ellipse selection by circle projection.

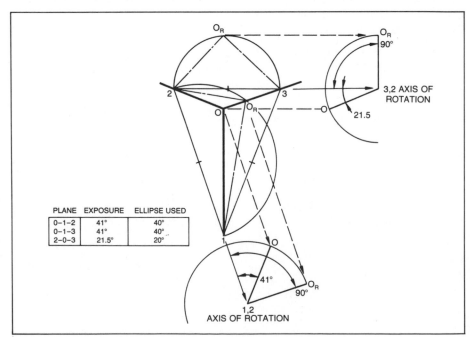

PLANE	EXPOSURE	ELLIPSE USED
0–1–2	41°	40°
0–1–3	41°	40°
2–0–3	21.5°	20°

Figure 5–16. Determining proper ellipse for axonometric face.

position in an auxiliary view. Remember that O is being revolved through a circular arc, which allows you to construct both positions of point O in the auxiliary view. Measure the angle between the line of sight (perpendicular to the axis of rotation) and the nonrevolved position of the principal plane.

Note that the closest standard ellipse is chosen as the proper solution for the problem.

SETTING UP A TRIMETRIC PROJECTION

Figure 5-17 shows the steps necessary to construct a trimetric projection. In Figure 5-17(a) you are given the scale detail drawings of a drill positioner. In Figure 5-17(b) the axes are chosen that give the desired object orientation. Next, an axonometric diagram is completed. Choose auxiliary views along the axes of rotation. Determine the closest standard ellipse (Figure 5-17c).

Orient the detail views for projection so that they are parallel to the edges of the *revolved* plane for that view. These revolved views often have to be moved away from the axonometric diagram to make room for the projection. Figure 5-17(d) shows the orien-

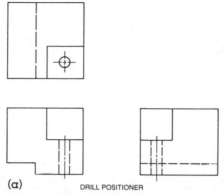

(a)   DRILL POSITIONER

Figure 5–17. Steps in setting up a trimetric projection.

tation of two principal views and the projection lines, which are perpendicular to the axis of rotation for that view.

Assemble the object by projecting the overall volume that encloses the drill positioner. Follow this with a systematic cutting of the block to reveal its final shape. Locate the center of the hole and two radius points on the centerlines. From your axonometric diagram (Figure 5-17c), you have determined that horizontal faces will have 30° ellipses. Fit a 30° ellipse on the centerlines to run through the two radius points.

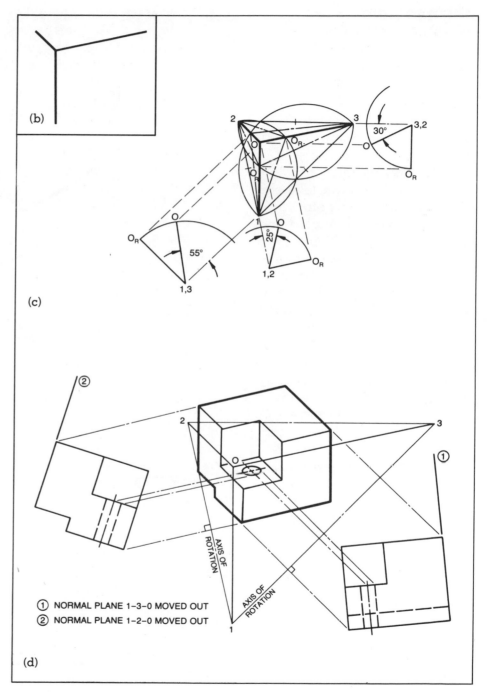

(b)

(c)

(d)

① NORMAL PLANE 1–3–0 MOVED OUT
② NORMAL PLANE 1–2–0 MOVED OUT

CONSTRUCTING NONAXONOMETRIC LINES

Lines that are parallel to the axonometric axes—*axonometric lines*—can be directly measured. Lines not parallel to the axes—*nonaxonometric lines*—cannot be directly measured.

In making a projection, nonaxonometric lines take care of themselves. End points are connected and the angled lines are found by default. In direct construction, the best way to draw nonaxonometric lines is to "box out" the object in a form easy to transfer into pictorial. Figure 5-18 shows an object enclosed in an isometric box. Four of the object's points rest on the perimeter of the form, making them easy to find. The ends of the nonisometric lines are connected to form the lines. Your knowledge that parallel lines will remain parallel in axonometric (and, of course, in this isometric example) provides a check on the construction and also shortens the time required to complete the drawing.

Figure 5–18. Nonaxonometric lines.

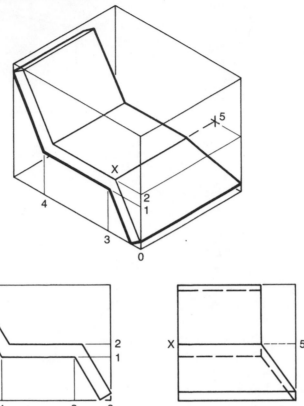

CONSTRUCTING ANGLES IN AXONOMETRIC

Figure 5-19 shows how angles are constructed on an isometric face. The same procedure would apply for angles on dimetric or trimetric faces.

The dimensions for a block have been given, allowing the overall form to be constructed.

What has *not* been given is the distance from the left side to the top of the incline. By constructing angle Ø at point 3, this distance can be determined. Two quick methods are available to the illustrator for constructing this angle:

1. *Laying off the legs or components of the angle.* Start with the bottom of the incline (point 3) and lay off distance X. Erect a vertical at this point. Distance Y will coincide with the top of the box. Connect points 3 and 4 to complete the angle.

2. *Making a circle projection.* Construct a fairly large ellipse of the correct exposure, using point 3 as the center. Project a circle whose diameter is equal to the major axis of the ellipse along the extension of the minor axis. Project point 1 onto the circle. Line 1-3 becomes the base line for angle Ø. Project point 2 back to the ellipse. Direct the angle on the axonometric face from point 3, through point 2, to the top of the box.

As you can see, both methods produce the same solution. Another method available to the illustrator is using an ellipse protractor. It is used like any protractor, with the angle above the baseline drawn from the center of the protractor to the appropriate tick-mark in the protractor's circumference. Since this is an *elliptical* protractor, the construction is done directly in axonometric. It does not require scale orthographic views or an auxiliary circle projection.

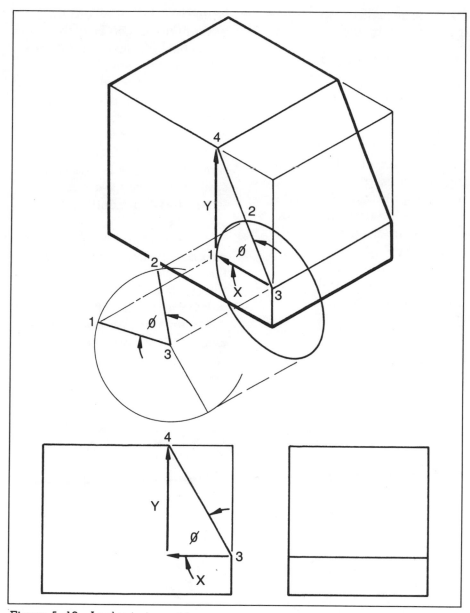

Figure 5–19. Angles in isometric.

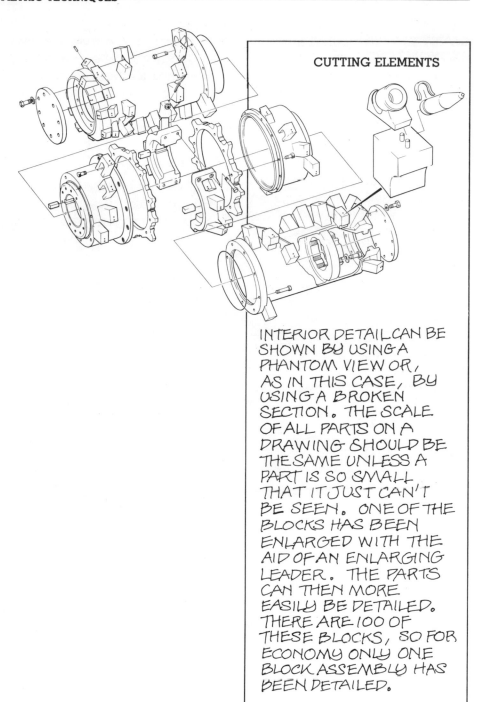

CUTTING ELEMENTS

INTERIOR DETAIL CAN BE SHOWN BY USING A PHANTOM VIEW OR, AS IN THIS CASE, BY USING A BROKEN SECTION. THE SCALE OF ALL PARTS ON A DRAWING SHOULD BE THE SAME UNLESS A PART IS SO SMALL THAT IT JUST CAN'T BE SEEN. ONE OF THE BLOCKS HAS BEEN ENLARGED WITH THE AID OF AN ENLARGING LEADER. THE PARTS CAN THEN MORE EASILY BE DETAILED. THERE ARE 100 OF THESE BLOCKS, SO FOR ECONOMY ONLY ONE BLOCK ASSEMBLY HAS BEEN DETAILED.

CONSTRUCTING CURVES IN AXONOMETRIC

When constructing curves that are not readily circular on axonometric planes, you can use a coordinate grid, as shown in Figure 5-20. You can find each point on the curve by a form of offset construction; each point is located back, over, and up from a known reference point.

To construct a curve, enclose the orthographic views in a box, and reproduce the box at scale in pictorial. Break up the curve into a number of points, and contain these points in a grid. Transfer the grid into pictorial. Likely points for grid lines are:

1. Points that are the farthest in or out on the curve.
2. Points of tangency between curve segments or points where the curve changes direction.

Since the grid lines are axonometric, the distance from the edge of the box to the curve can be directly measured. Connect the points with a smooth curve. The bottom of the object is a parallel curve the thickness of the object away from the curve on the top. Either drop verticals from the top points or repeat the grid construction on the bottom of the box.

This technique is helpful when doing sectional construction, as was described in Chapter Four.

CONSTRUCTING CIRCLES IN AXONOMETRIC

When constructing circles, you should use an ellipse template whenever possible—even

when you have to "put together" an ellipse from the parts of several others (as I will describe later). Still, there are situations in which you have to construct an ellipse. An off-exposure, odd-size, or oversize ellipse will probably require construction. There are several methods available; the two presented here are the easiest to learn and the most rapid to complete.

In the *grid method*, a circle can be enclosed in a square, the sides of which are equal to the circle's diameter. The circle will be tangent to the square at the center or mid-point of each side. Both the orthographic and isometric views of the enclosing box are shown in Figure 5-21.

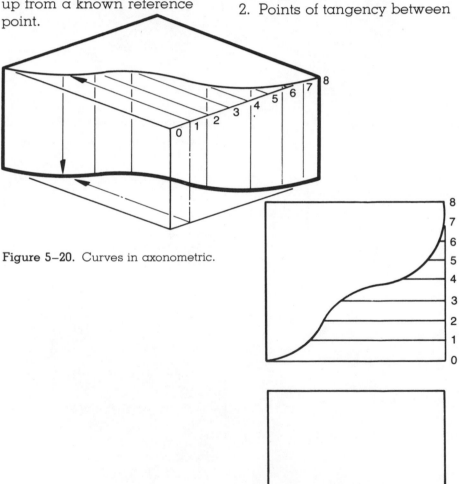

Figure 5-20. Curves in axonometric.

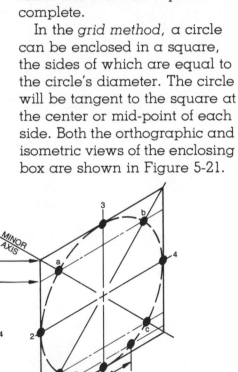

Figure 5-21. Grid method of constructing an ellipse.

Figure 5–22. Constructing an ellipse—four center method.

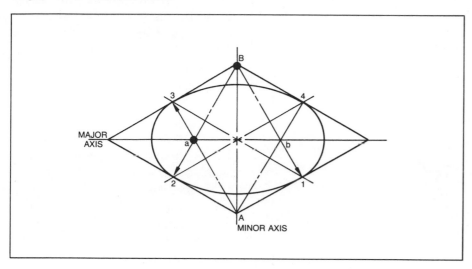

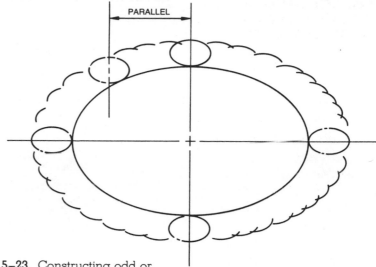

Figure 5–23. Constructing odd or oversize ellipses.

The circle has been inscribed within the orthographic box, and the mid-points—1, 2, 3, and 4—have been identified. Furthermore, the position of the circle on the diagonals has been marked. Note how these points have been transferred into isometric. Making use of the symmetry of the circle, two intermediate points have been located by making only one measurement, distance A. In fact, the corresponding two points above can be found by projecting the grid and distance A to the top. The ellipse is completed with a curve through the points.

The second method, shown in Figure 5-22, is called *four-center approximation.* Two pairs of arcs are used to rough out an ellipse. It looks like a fat hot dog, but it is accurate enough for most illustration.

In order to use this method, construct an axonometric box of equal sides. At the centers of the sides—1, 2, 3, and 4—construct perpendiculars. The perpendiculars from 1 and 2 will meet on the minor axis at point B. The perpendiculars from 3 and 4 will meet on the minor axis at A. In isometric drawing, A and B will coincide with the ends of the minor axis. In dimetric ellipses, the lines will still meet on the minor axis but not at its ends.

Perpendiculars 2-3 and 1-4 will cross on the major axis at a and b, respectively. Points a and b will be on the major axis in any case. Swing arcs B-2 and A-3; swing arcs a-2 and b-1. These arcs should be tangent at points 1, 2, 3, and 4.

For the four-center method to work, the enclosing form must be an equal-sided parallelogram or rhombus. For each dimetric situation, one of the three faces will have equal scales, and a four-center method will work on this face. Of course, this is not the case in trimetric, where no face has two equal scales.

CONSTRUCTING ODD-SIZE OR OVERSIZE ELLIPSES

Ellipse guides offer a selection of ellipses in common size increments. To fill the gaps in between, ellipses can be used in combination, as shown in Figure 5-23.

The smaller ellipse rotates around the larger, with its minor axis constantly parallel to that of the larger. The newly plotted ellipse is described by the outside traces. Use an adjustable curve to smooth this shape.

USING CONCENTRIC ELLIPSES AS A WAY OF MEASURING NONAXONOMETRIC LINES

Because the axonometric view of a circle, like the normal view of any circle, has uniform radii, concentric circles or ellipses can be used to measure distances along lines that are not parallel to the axonometric axes.

Once the desired line has been laid off (using one of the axonometric protractors found in Appendix D), the line can be measured by underlaying a guide, as has been done in Figure 5-24. If the line length falls between the common values, you can eyeball the difference.

It would be a good idea to produce your own concentric measuring scales, one for each of the available ellipse guides. With scales like that in Figure 5-25, you can measure almost any line, in any plane, in any axonometric situation.

LETTERING ON AXONOMETRIC DRAWINGS

Pictorial drawings require numbers and text (call-outs) at times. Call-outs can be added by a variety of means, following either of these two schemes:

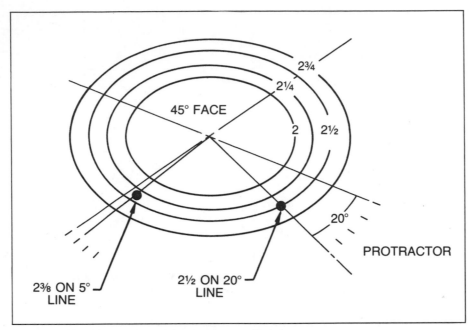

Figure 5–24. Measuring with concentric ellipses.

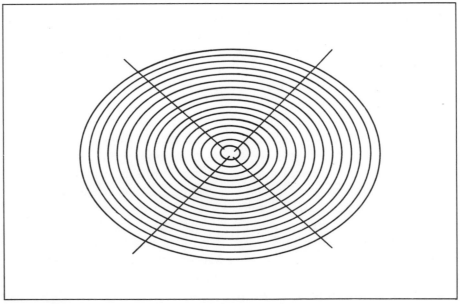

Figure 5–25. 45° concentric ellipse scale—4″ to ¼″ in ¼″ diametric steps.

1. *In the plane of the paper.* As shown in Figure 5-26, all text done in this manner is off of the object, on the picture plane. This method makes it easy to use transfer type, Leroy lettering, or photo-type.

2. *In the plane of the drawing.* This approach is more difficult—even impossible at times. Some mechanical lettering devices can be adjusted to produce inclined letters. Photo-type can be manipulated. In the case of computer graphics, type on any face is possible. But for the most part, since strike-on and transfer type are generally used, placing type on the planes of the drawing is best suited to hand-lettered text. Some recent mechanical scribers can be adjusted to accommodate most axonometric faces.

DETERMINING THE SCALE FOR AN AXONOMETRIC DRAWING

As mentioned earlier in this chapter, you should make an illustration at the largest comfortable scale. This often means using an odd drawing scale, which is not a great problem. The steps for setting up a scale of a cylindrical object are shown in Figure 5-27.

1. After choosing the view and orientation, determine the largest circular diameter used on the object.

2. In the proper orientation and using the correct ellipse exposure, lay out a 4″ ellipse—normally the largest complete ellipse generally available.

3. Knowing the radius of the object, divide the vertical and horizontal radii into the same number of divisions, using the method shown.

4. Transfer this scale to axes moved away from the drawing area. You can then use these scales to measure dimensions on the illustration.

DETERMINING THE SCALE OF AN ORTHOGRAPHIC DRAWING

Many engineering drawings are reproduced at less than 100% for ease of handling. When this is done, seldom is the reproduction percentage noted. In such cases the drawing can't be directly scaled. Furthermore, many drawings are drawn not to scale (NTS). When this happens only the numerical values can be used. But if the drawing was originally made at scale, you can construct a new scale right on the drawing so that all distances can be determined.

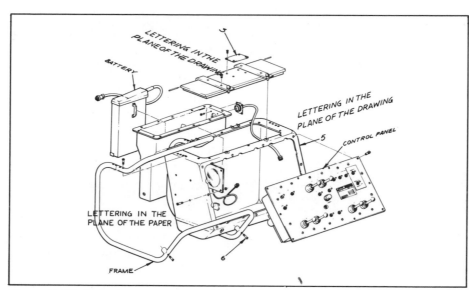

Figure 5–26. Lettering on axonometric drawings.

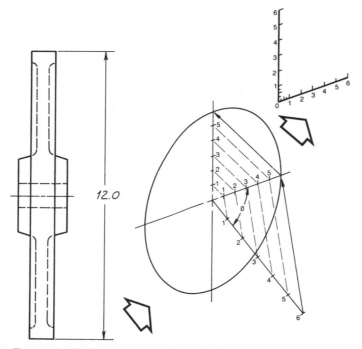

Figure 5–27. Determining the scale for an axonometric drawing.

As shown in Figure 5-28 (page 94), the procedure is:

1. Locate a value of suitable length. If there is one that is a whole number, use it.

2. Lay this distance off on a clear space on the drawing.

3. Construct a guide line from one end, marking off an equal number of divisions as the original distance.

4. From the last division, connect to the end of the original line.

5. Construct parallels, dividing the original line.

6. This is the scale for the drawing. Compare distances on the drawing to this scale.

ENGINEERING DRAWING

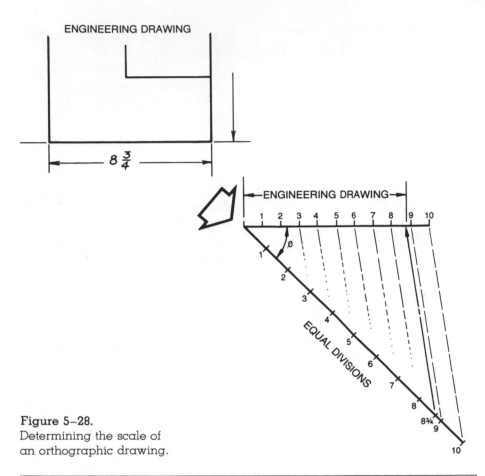

$8 \frac{3}{4}$

ENGINEERING DRAWING

EQUAL DIVISIONS

Figure 5–28.
Determining the scale of
an orthographic drawing.

List of Terms

Many of these terms are used by illustrators to describe what they do. You shold be able to recognize them and in time use them to describe *your* illustration.

axonometric diagram
axonometric drawing
axonometric grids
axonometric projection
cardinal view
circle projection
dimetric view

ellipse protractor
ellipse selector
exact illustration
fudge factor
intended use
interpolation

isometric view
principal orthographic view
representative illustration
reproduction
resources
standard axonometric view
trimetric view

Problems for Further Study

Difficult objects are made up of several not-so-difficult-to-construct shapes. Learn to construct the basic forms—prism, cylinder, cone, sphere, pyramid, and torus. When used together, these and a few other shapes can make up almost any object.

The following problems give you the opportunity to practice dimetric and trimetric drawing.

In some cases, axonometric axes have been given and you are expected to make a traditional axonometric projection. Other problems suggest an axonometric orientation for direct construction. A few problems have been included without suggestions, allowing you and your instructor to determine the appropriate orientation.

1–6. Draw 20°–40° dimetric views of these objects. Orient the longest face or the most detail in the long or 40° face. Don't forget that alternate views exist with 40° or 20° on the bottom plane, resulting in an upward look.

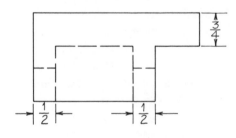

$\frac{3}{4}$

$\frac{1}{2}$

$\frac{1}{2}$

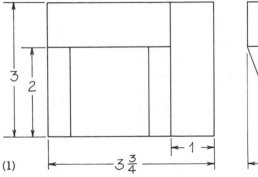

3

2

$3 \frac{3}{4}$

1

(1)

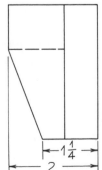

$1 \frac{1}{4}$

2

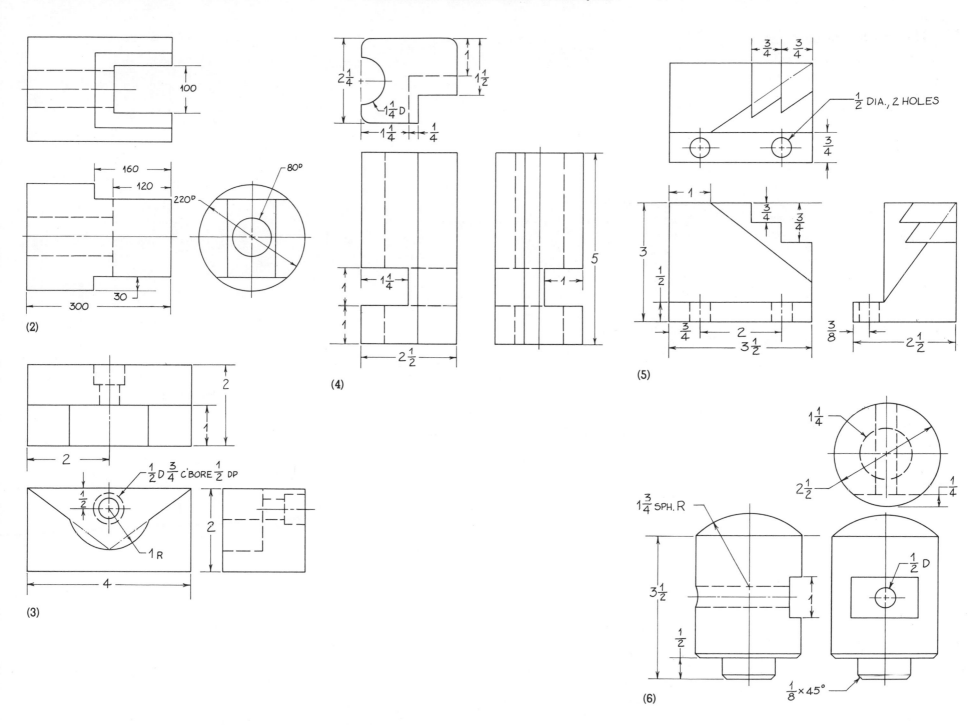

(2)

(3)

(4)

(5)

(6)

7–10. Draw 15°–40°–50° trimetric views of these objects. Pick an orientation that best shows the features of each, keeping the differing exposure of each face in mind.

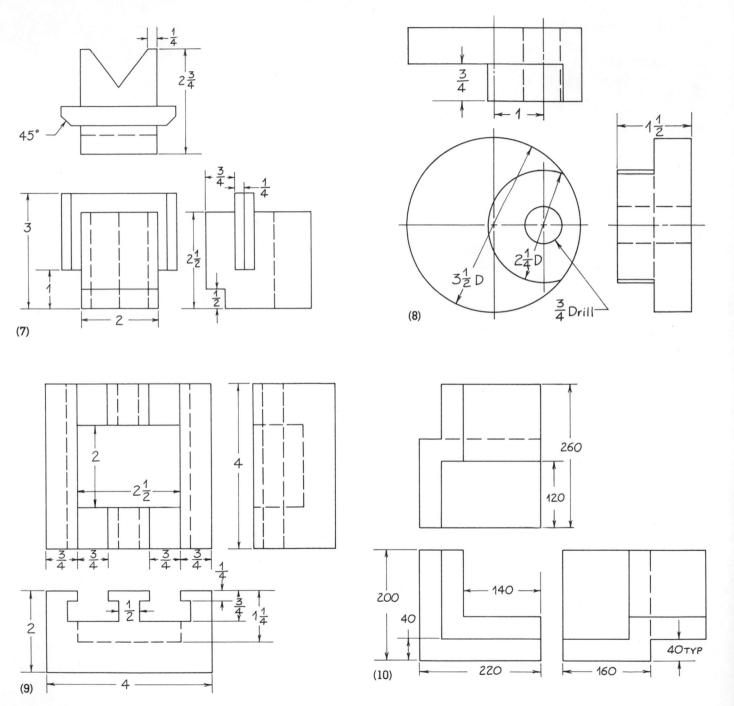

(7)

(8)

(9)

(10)

11–13. Choose a dimetric or trimetric view appropriate for each object. Note the size of the parts and adjust your scale up or down. Consult Tables C-5 and C-6 in the Appendix for dimetric and trimetric information.

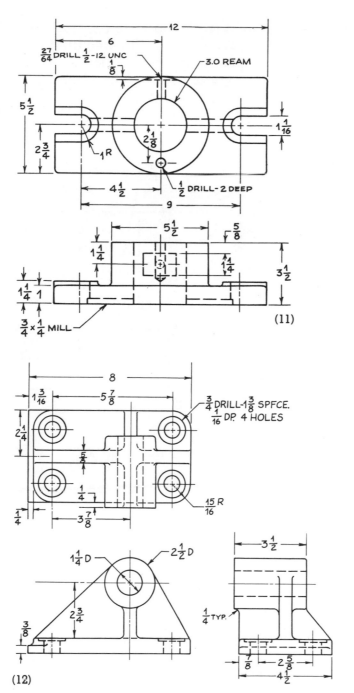

(11)

(12)

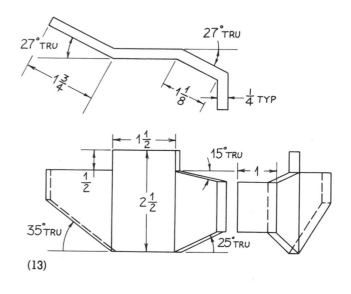

(13)

14. Given a scale drawing of the object, produce an axonometric drawing by directly scaling the views.

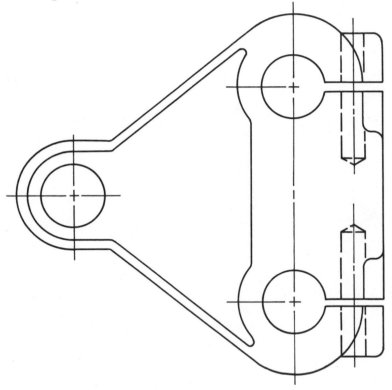

SCALE: HALF

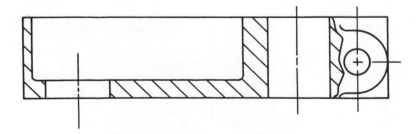

15. Produce an exploded axonometric drawing of the assembly shown. Note the information given in the description of parts.

1. BASE
2. PIN, 1 D × 4¾ - ⅛ ×45° CHAMFERS, ⅛ HOLE
3. ROLLER, 1 I.D × 2½ O.D, ¼ ×1 RELIEF
4. COLLAR, 1 I.D × 1½ O.D - 2 RQD.
5. WASHER, 1 I.D × 1½ O.D.
6. PIN, COTTER ⅛ × 1½

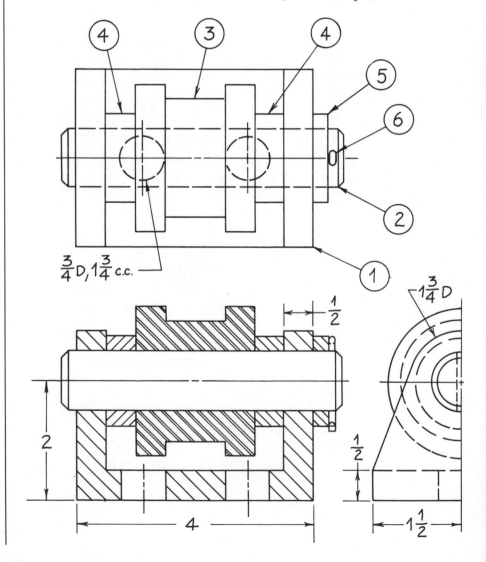

16–17. Construct 25°–25°–50° dimetric views of the objects shown. Problem 16 should be drawn at full scale. Problem 17 should be drawn twice size.

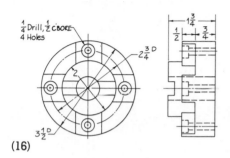

(16)

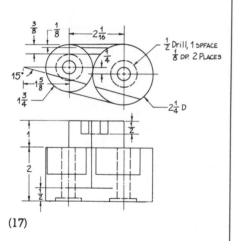

(17)

18. Make an exploded assembly drawing of the object shown.

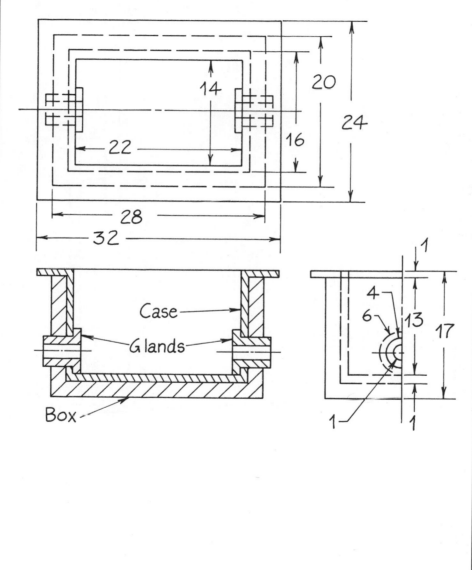

Case

Glands

Box

19–21. Make a 50°–30°–25° trimetric drawing of each of the objects.

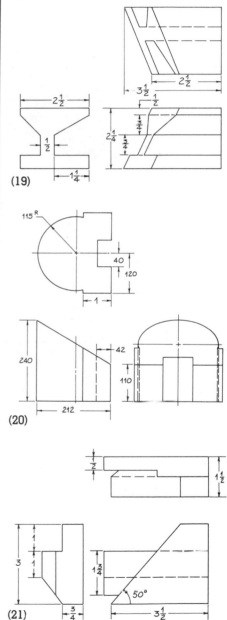

(19)

(20)

(21)

22. Make an axonometric exploded assembly drawing at half scale.

23. Scale the views of the link and produce an axonometric drawing, using either dimetric or trimetric.

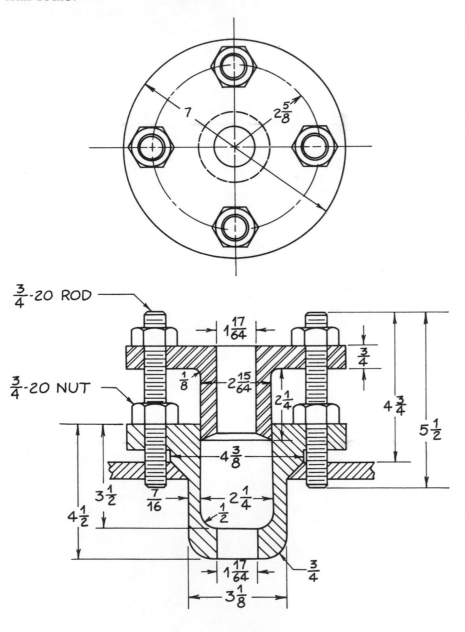

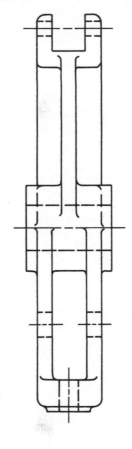

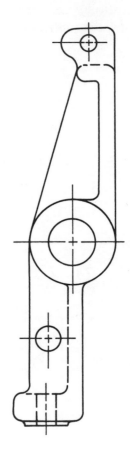

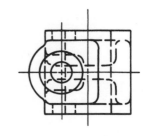

24. Choose a view and orientation that reveals the most of the object's features. Try to show the spotfaced holes, the curved cut, and the curved rib.

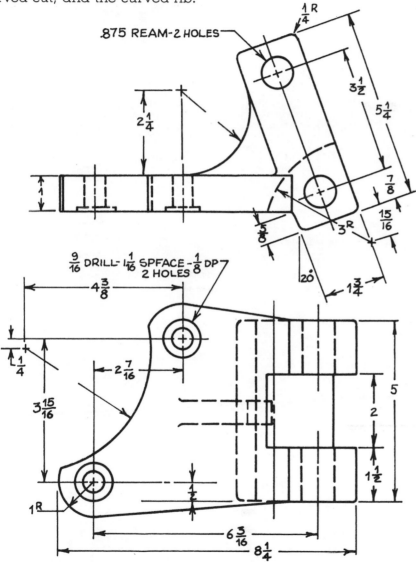

CHAPTER SIX Perspective Drawing

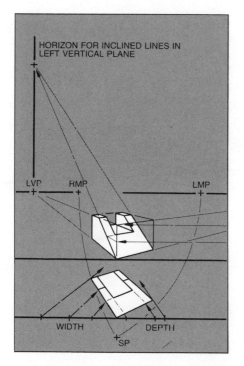

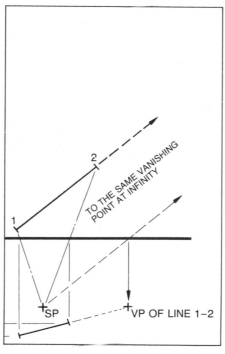

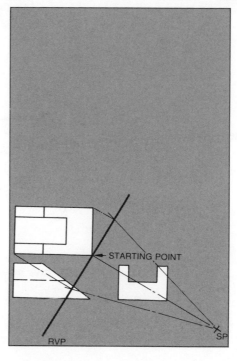

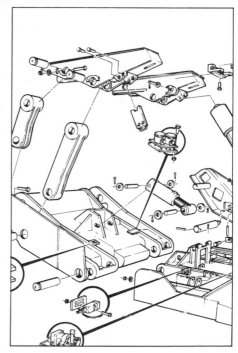

To many people, perspectives are *any* pictorial drawings. But as we have seen, some pictorials are axonometric or parallel drawings. Axonometric views are orthographic views—parallel lines remain parallel. In perspective, parallel lines remain parallel in reality but on the drawing they visually converge at a distant point. So when someone asks you to do a "perspective" of an object, be careful! He or she probably means a *pictorial* view, not knowing the difference between axonometric and perspective drawing.

Trimetric drawing is similar to perspective drawing in that both produce a realistic view. Perspective drawing is really no more difficult than axonometric drawing; it's just different.

Drawing vs. Projection

Just as was the case in axonometric drawing, there are basically two methods of producing perspective drawing: projection and direct construction. The projection method is called *visual ray perspective*; the construction method is called *measuring point per-*

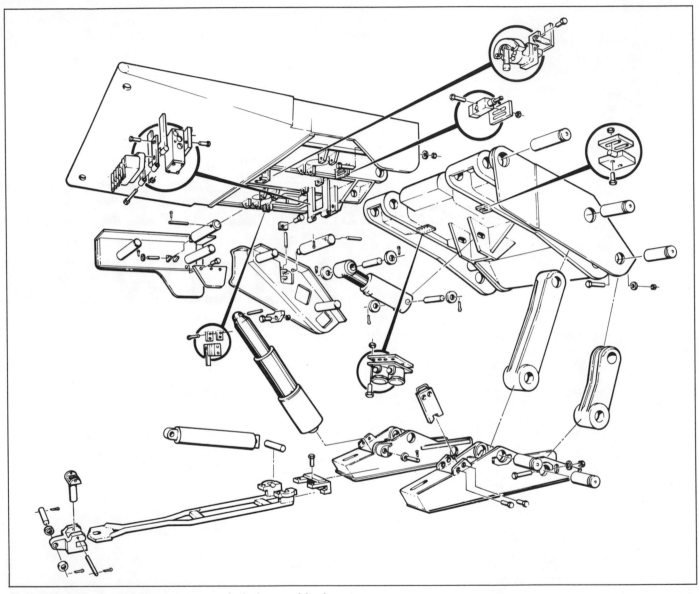

Figure 6–1. Technical perspective exploded assembly drawing.

spective. Of the two, measuring point perspective is the more powerful tool in terms of getting the illustrator "into" the drawing. Some illustrators feel that technically accurate perspective is beyond their abilities because of perspective drawing's dependence on involved and laborious construction.

In order to have the confidence to set up your own perspective, starting with a clean sheet of paper, you must understand the relationship of the components of perspective drawing. Once you understand them, you will be best served to use projection theory *only* to set up the perspective. Once you have the perspective set up, you should directly construct the object rather than project it. Again, you must think of the drawings as three-dimensional. If you don't, perspective will forever be a chore.

Because perspective drawing is less important than axonometric drawing in technical illustration, only the basics of perspective drawing are in this chapter. After completing this material you may want to pursue the topic by looking into any of the several fine perspective texts on the market.

There are several shortcuts to drawing perspective, some of which are presented in this chapter; you will find that most perspective texts make use of them. The ability to effectively use these shortcuts depends on your knowledge of the foundations of perspective. The real mechanics of what happens between object, viewer, and picture plane must be understood first, with the tricks coming later. For these reasons I will present perspective first in its fundamental projection. Once you understand fundamental projection, you can master the shortcut methods, eventually arriving at a technique of perspective drawing that is almost as easy as isometric drawing.

The Basics of Perspective

The perspective image is created by projecting an object onto a picture plane. Instead of using a viewing point set at infinity (yielding parallel lines of sight and an orthographic view), perspective sets the viewer at a known distance from the object, causing lines of sight to converge at the observer.

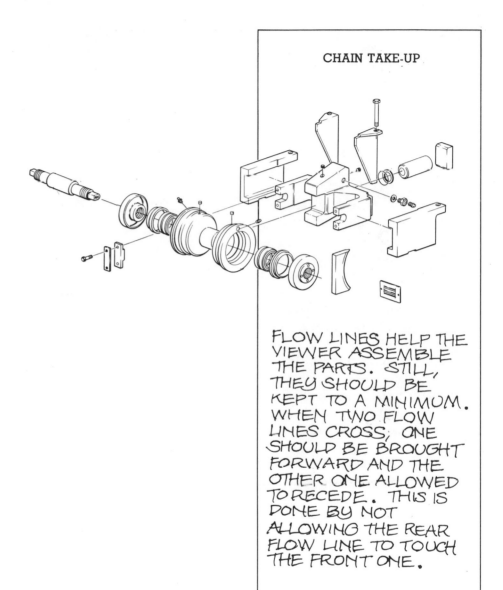

CHAIN TAKE-UP

FLOW LINES HELP THE VIEWER ASSEMBLE THE PARTS. STILL, THEY SHOULD BE KEPT TO A MINIMUM. WHEN TWO FLOW LINES CROSS, ONE SHOULD BE BROUGHT FORWARD AND THE OTHER ONE ALLOWED TO RECEDE. THIS IS DONE BY NOT ALLOWING THE REAR FLOW LINE TO TOUCH THE FRONT ONE.

There are several terms unique to perspective drawing that you should become familiar with. They include:

Station point (SP). Location of the observer.

Picture plane (PP). Plane upon which the perspective view is drawn.

Horizon line (HL). Line on which exist all points where lines parallel to the horizon are directed.

Ground line (GL). Intersection of the picture plane and the ground.

Vanishing point (VP). Point at infinity to which parallel lines will be directed.

Visual ray. Line of sight from the observer.

Measuring wall. Wall built from the object to the picture plane in order to measure height.

Measuring line (ML). Intersection of the measuring wall and the picture plane.

Measuring point (MP). A vanishing point used to determine diagonals and depth measurement.

Horizontal measuring line (HML). Arbitrary measuring line for use with measuring points.

THE OBSERVER

Of course *you* are the observer in perspective drawing. Your direction of sight is perpendicular to the picture plane, with your eyes picking up everything within your *cone of vision* (Figure 6-2). This cone is roughly elliptical and is differ-

ent for each person. The further from the center of vision a point falls within the cone, the greater the distortion. To keep the distortion within acceptable limits, a 30° plan and 15° elevation cone may be used.

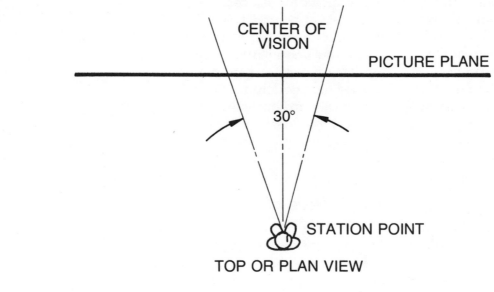

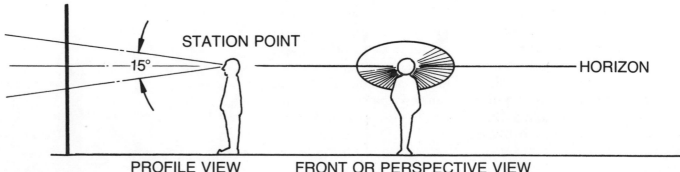

Figure 6–2. The cone of vision.

THE OBJECT

The relationship of the object, picture plane, and observer determines the nature of the perspective drawing. If the picture plane is brought too close to the observer, the size of the drawing is greatly reduced. Even so, it is better to move the picture plane than to move the object. Otherwise the object may fall outside the cone of vision. Figure 6-3 shows this relationship of observer, picture plane, and object. If the object touches the picture plane, that point will appear true height on the drawing. If the point doesn't lie on the picture plane, a measuring wall must be built to touch the picture plane. As you can see in the example, the observer sees everything in front of him projected forward to the picture plane.

If the object is set at an angle to the picture plane, the perspective is said to be *angular*. If the object is perpendicular and parallel to the picture plane, the perspective is said to be *parallel*. If the object is angular and tipped forward or back, the perspective is said to be *oblique*. These types of perspective are sometimes incorrectly called one-point, two-point, and three-point perspectives.

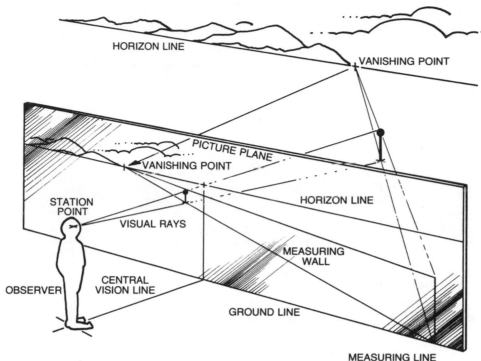

Figure 6–3. Perspective theory.

Figure 6–4. Perspective by fundamental projection. (a) Perspective from profile view and top view.

FUNDAMENTAL PROJECTION

The perspective drawing is formed by the intersection of lines of sight from the observer (visual rays) and a plane (picture plane). Figure 6-4(a) shows the observer standing in front of the picture plane looking at an object on the ground behind the picture plane. Note that the intersections of the visual rays in the *profile view* provide heights for the perspective (front) view. The intersections of the visual rays and the picture plane in the *top view* provide widths for the perspective view. The object is assembled by normal orthographic projection.

The reason that this method isn't used other than to demonstrate how the perspective image is projected is because the profile view must appear in a nonprincipal orientation, which usually means redrawing the view. To make perspective drawing more efficient, a simplified method of projection is used, one that does not require projecting visual ray—picture plane intersections from the profile view.

Several attributes of projected perspective have been standardized into construction methods designed to increase efficiency in perspective drawing. These standardized methods keep you from having to redraw the side view and continuously make visual projections. These methods include:

1. That any point on the picture plane will be seen at its true height in perspective. Knowing this, the height of a point not on the picture plane can be measured when another point of the same height is on the picture plane. *This is the basis for building measuring walls and measuring lines* (Figure 6-4b).

2. That parallel lines will visually meet at a common point in the distance. *This is known as a vanishing point* (Figure 6-4c).

3. That the vanishing points of all lines in parallel planes will define a line. *This line is the horizon line for those planes* (Figure 6-4d).

4. That since a vanishing point is really at infinity, a visual ray parallel to a line will be directed to the line's vanishing point. Where this ray pierces the picture plane is the perspective view of the line's vanishing point. *This allows the vanishing point for any perspective line to be found* (Figure 6-4e).

If you know this, you can draw any perspective: If you can locate a point on, in front of, or behind the picture plane, you can draw any perspective be-

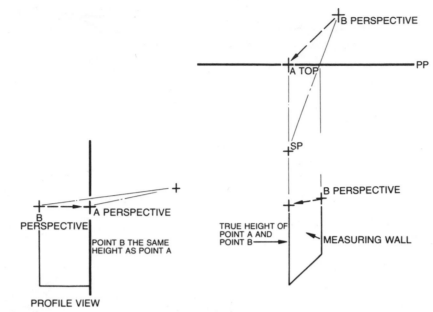

Figure 6–4. (b) Building a measuring wall.

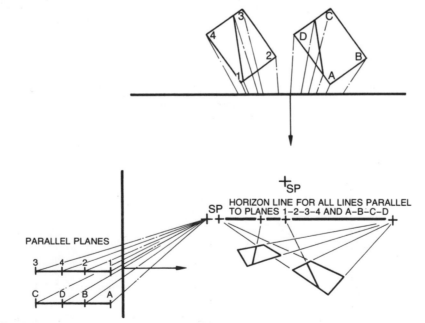

Figure 6–4. (c) Parallel lines converge at a common point.

cause any object, no matter how difficult, can be described as a group of points. In Figure 6-5 these points have been identified in the top and side views. We can now make use of vanishing points, measuring walls, measuring lines, and the horizon line to construct a perspective drawing of the three points. The height information from the profile or side view does not have to be projected; rather, it can be measured with a scale or transferred with dividers from a separate orthographic view.

Point A exists on the picture plane and will be true height. Point B is behind the picture plane; it must be projected forward to the picture plane to find its true height. Point C is in front of the picture plane. It must be projected back to the picture plane to measure its true height.

At what angle should you build the measuring walls for determining these heights? Pick a convenient vanishing point, say VP_1. Go perpendicular back to the picture plane from VP_1 and then to the station point. This is the angle that all measuring walls vanishing to VP_1 must be built parallel to.

The true height of point A can be brought over from its position in the side view. For point B,

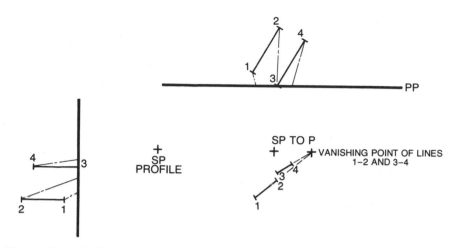

Figure 6–4. (d) Vanishing points of lines in parallel planes form a horizon line.

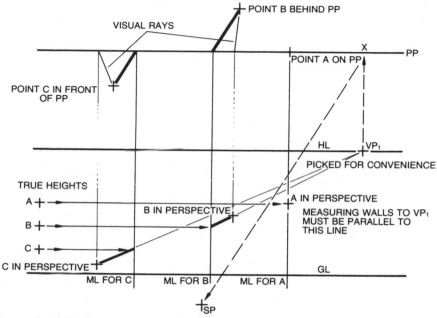

Figure 6–5. Finding the three types of points.

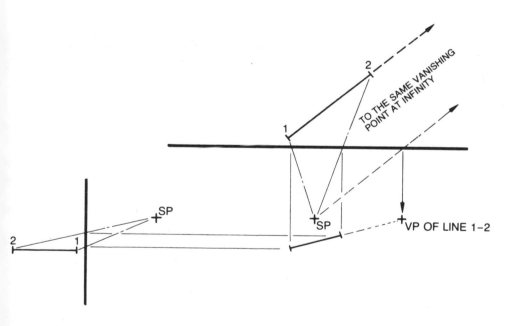

Figure 6–4. (e) Finding the vanishing point for a line.

build a measuring wall forward to the picture plane parallel to the SP-X line. Build a measuring wall back to the picture plane parallel to SP–X for point C.

Where these measuring walls intersect the picture plane, bring down measuring lines. True heights are brought across to the appropriate measuring lines. Now be careful! The true height line *goes back* to find point B; it *comes forward* to find point C.

To locate where points B and C lie, on the top of the two measuring walls, take a line of sight from the SP to the point in question. Where this line of sight pierces the picture plane, project straight down until this meets the true height line going back in perspective. This intersection is the location of the point in perspective.

To review, any point in the perspective drawing must be found two ways:

1. By the true height of the point, as measured on a measuring line on the picture plane.

2. By the location of the point on the picture plane, as found by a line of sight to the point piercing the picture plane.

109

Setting up a Perspective

By far the most difficult part of perspective drawing is getting the drawing started. This involves both decision and mechanics.

DECISIONS

1. How should the object be set in relation to the picture plane?
2. How far away is the object from the observer?
3. How high above or below the HL is the GL?

It is best to place the observer in the middle of the object so that the object fits into a 30° cone of vision. This position places the observer the closest to the object without causing unacceptable distortion.

The lowest that the GL can be below the HL is the point where the front corner of the object is visually 90°. The visual angle a 90° corner assumes in perspective is from 90° to 180°. This corner quite literally cannot be less than 90° without distortion. This same rule applies to placing the GL above the HL so that the observer is looking up at the object. Between these two extremes, it is the illustrator's choice of top or bottom exposure (Figure 6-6b).

Distortion begins as you move from the center of vision and increases as you move farther away. A general guideline is to place the object in the middle of the cone, balancing the distortion around the object. As you can see in Figure 6-6(a), positioning the object to the side of the cone of vision requires the viewer to back away from the picture plane. One positive result of backing the viewer away from the picture plane is to approach parallel lines of sight, making the perspective seem less drastic. To accomplish both ends—reducing distortion and the drastic or rapidly vanishing look of some perspectives—keep the viewer as far away from the picture plane as possible, with the object in the center of the cone of vision.

MECHANICS

Figure 6-7 is provided to show the steps in setting up a visual ray perspective. The object is placed touching the picture plane with its faces at 30° and 60°. This is called a 30°–60° perspective. A 30°–60° perspective is easy to draw because of common angle projection. It is popular because it produces one shallow and one fully exposed vertical plane.

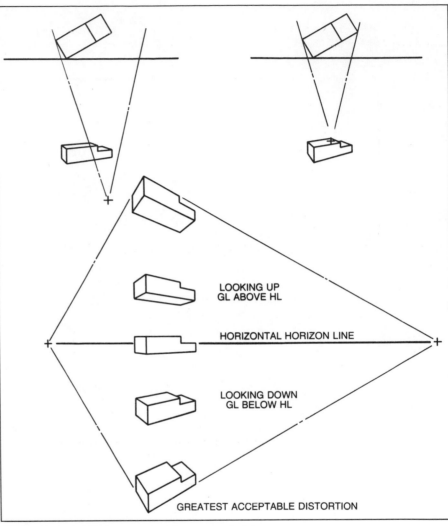

LOOKING UP
GL ABOVE HL

HORIZONTAL HORIZON LINE

LOOKING DOWN
GL BELOW HL

GREATEST ACCEPTABLE DISTORTION

Figure 6-6. (a) Positioning in the cone of vision. (b) Ground and horizon lines in distortion.

1. Draw a horizontal line (PP) and place the top view of the subject so that it touches the PP at the front corner and so that the sides are inclined to the PP at 30° and 60° (Figure 6-7a).

2. Draw perpendiculars from the corners of your object. Bisect this distance (Figure 6-7b).

3. Place the station point (SP) along a line from the center of this distance so that the object is within a 30° cone (Figure 6-7c).

4. Draw the horizon line and ground line parallel to the PP. The HL to PP distance is unimportant. The HL to GL distance controls downward or upward orientation (Figure 6-7d).

5. From the SP, project lines to the PP parallel to the object's sides. Drop points x and y straight to the HL. These are the vanishing points to which the right and left sides of your object vanish (Figure 6-7d).

6. Build the form of the object using the point where it touches the PP as the starting point and true height measuring line (Figure 6-7e).

7. Cut the slot out of the top of the block. Measure how far the cut extends into the block at the front corner and take that true height back along the left side until it intersects the two visual ray lines at the slot's edges (Figure 6-7f).

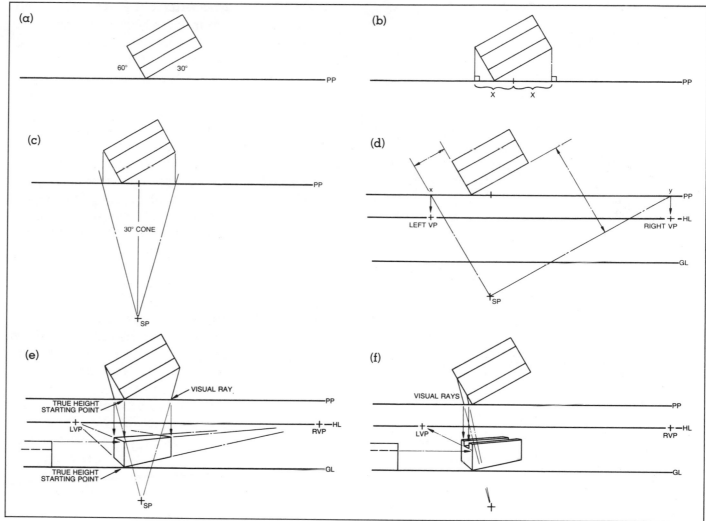

Figure 6–7. (a) Setting up the perspective.
(b) Finding center of object.
(c) Finding station point.
(d) Finding vanishing points, horizon, and ground.
(e) Building the form.
(f) Finishing the perspective.

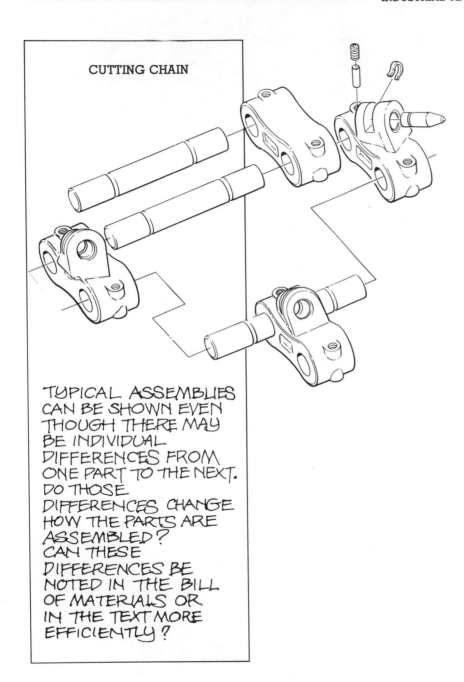

CUTTING CHAIN

TYPICAL ASSEMBLIES
CAN BE SHOWN EVEN
THOUGH THERE MAY
BE INDIVIDUAL
DIFFERENCES FROM
ONE PART TO THE NEXT.
DO THOSE
DIFFERENCES CHANGE
HOW THE PARTS ARE
ASSEMBLED?
CAN THESE
DIFFERENCES BE
NOTED IN THE BILL
OF MATERIALS OR
IN THE TEXT MORE
EFFICIENTLY?

Measuring Point Method

Perspectives can be drawn without having the top view in position on the PP and without taking continuous visual rays to locate each point on the drawing. Basically, this alternative method, measuring point perspective, involves creating what is called a *perspective plan* below the object.

Measuring point perspective further frees the illustrator from positioning orthographic views in the top and profile views, as in fundamental projection, or in the top view, as in visual ray perspective. Measuring point perspective is particularly effective when the scale drawings are very large, when drawings were not made to scale, and when no drawings exist and only the object's numerical dimensions are known.

This approach is an attempt to make perspective drawing more efficient and relies on the principles of fundamental perspective projection for its construction.

Measuring point perspective is based on the knowledge that if an object were to be pressed flat against the picture plane, measurements along that surface would be true size (Figure 6-8a). When an object is inclined to the picture plane, these measurements are rotated, as in Figure 6-8(b). If you could easily draw these rotations in perspective, you could determine their measurements. An easier method relies on the chords of these arcs. Since the chords are parallel, they will converge at a common vanishing point. Because this vanishing point is used to measure horizontal distances, it is called a measuring point (MP). Each line of a different inclination to the picture plane will have its own measuring point as determined by the vanishing point of its rotation arc chords. Rather than doing this rotation at the picture plane for every line, a construction technique is employed in which the distance from the line's vanishing point to the station point is rotated until it is in the picture plane. This point on the horizon line is the vanishing point of the parallel arc chords for that line. This is shown in Figure 6-8(c).

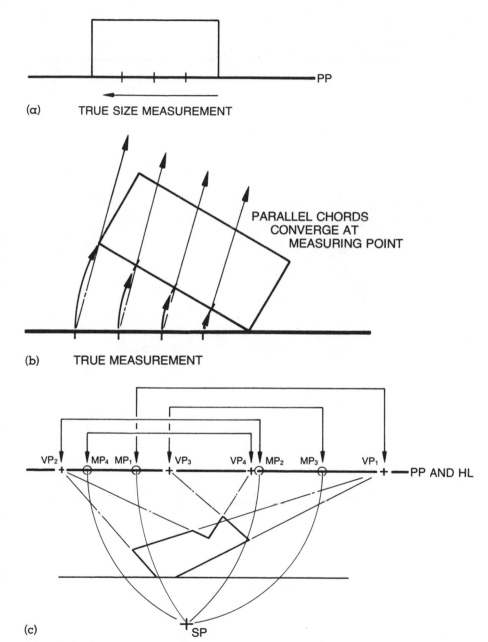

(a) TRUE SIZE MEASUREMENT

PARALLEL CHORDS CONVERGE AT MEASURING POINT

(b) TRUE MEASUREMENT

(c)

Figure 6–8. (a) Object in contact with picture plane. (b) Object not flush with picture plane. (c) Locating measuring points for different lines.

DECISIONS

Figure 6-9 will provide an example of measuring point perspective. Make a scale thumbnail sketch:

1. Incline the object to the PP as before.
2. Locate the SP in a 30° cone.
3. Determine the scale distance between VPs.

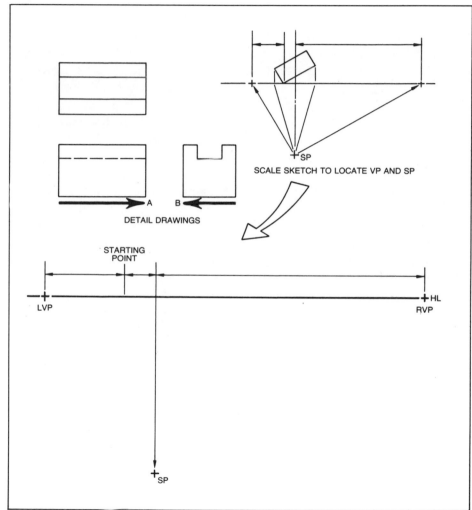

SCALE SKETCH TO LOCATE VP AND SP

DETAIL DRAWINGS

STARTING POINT

Figure 6–9. (a) Measuring point perspective setup.

MECHANICS

1. Draw the HL. Locate vanishing points at the scale at which you wish to produce your perspective by referring to the scale distance on your sketch (Figure 6-9a).

2. Rotate distance RVP–SP to find the right measuring point. Rotate distance LVP–SP to locate the left measuring point (Figure 6-9b).

3. At a convenient location above or below the GL, construct a horizontal measuring line. This is the starting location of your perspective plan. Locate where the object touches the PP in your scale sketch. Find this point at drawing scale on the HML (Figure 6-9c).

4. Lay out distance A to the right of the starting corner. Go to the RMP. The place where this line crosses the right base line represents distance A in perspective (Figure 6-9d).

5. Repeat for distance B. An alternative method for finding distance B is shown in the detail box. Since B describes a line going to the left vanishing point, a left starting point is needed.

Go from the end of distance A through the LMP to get a left starting point on the HML. Measure distance B to the left of the starting point. Go back to the LMP. This determines the same depth at the right of the object as was found with the first method on the left.

6. When the perspective plan is completed, raise the corners and, using the front corner as a true height measuring line, build the object. Measure the height of the slot by building a measuring wall to the PP. Bring up a measuring line. Take this true height back until it meets the corner coming up from the perspective plan (Figure 6-9e). Since in this case the object touches the picture plane at its front corner, the corner can be used as a measuring line for heights in either right or left vertical planes. This method is often called "chasing the heights" because the height lines are chased over and

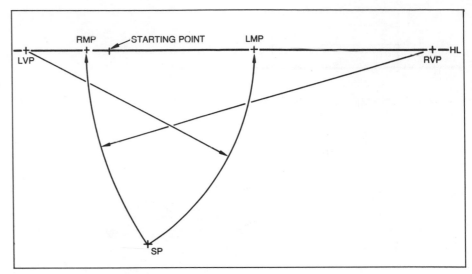

Figure 6–9. (b) Locating measuring points.

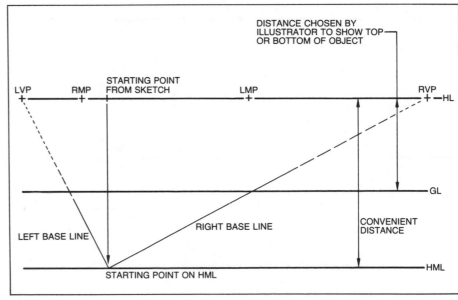

Figure 6–9. (c) Locating HML, GL, and starting point.

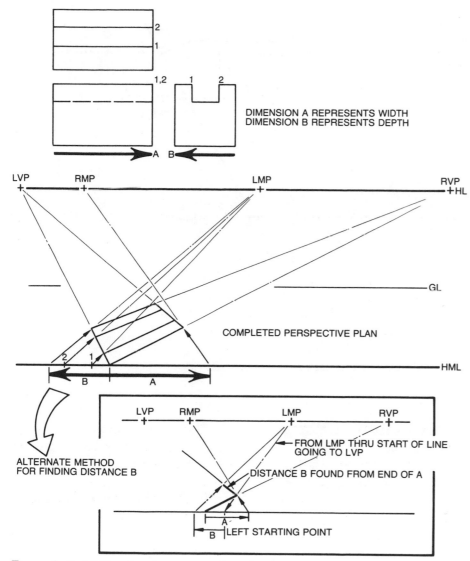

DIMENSION A REPRESENTS WIDTH
DIMENSION B REPRESENTS DEPTH

COMPLETED PERSPECTIVE PLAN

ALTERNATE METHOD
FOR FINDING DISTANCE B

LVP   RMP   LMP   RVP

←— FROM LMP THRU START OF LINE
      GOING TO LVP

DISTANCE B FOUND FROM END OF A

LEFT STARTING POINT

Figure 6–9. (d) Constructing perspective plan.

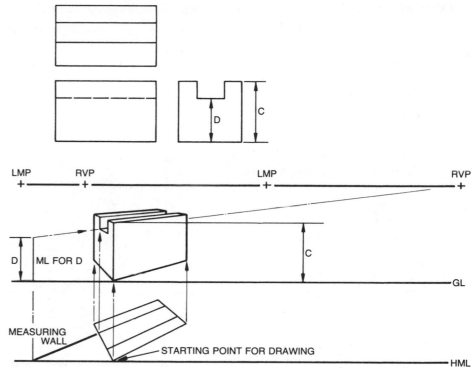

ML FOR D

MEASURING
WALL

STARTING POINT FOR DRAWING

Figure 6–9. (e) Finding heights and finishing the perspective figure.

across the object. But at each direction change, error can make the final point considerably off position. The singular measuring wall eliminates this error build-up. Using the front corner works only because it touches the picture plane. Building measuring walls works whether this special condition exists or not.

There are advantages to both perspective methods. The visual ray method produces a flat projection yet a fairly rapid drawing *if* you have the scale detail drawings. The measuring point method produces a perspective model. If you are doing several repetitious perspectives, the measuring point method is faster. If you are doing only one, the visual ray method is fine.

A TYPICAL PROBLEM

You are given the detail drawings of a wedge take-up block. You have decided to construct an angular perspective using the measuring point or perspective plan method. How would you go about it?

1. Do a thumbnail sketch (Figure 6-10) to determine how the block is to be inclined to the picture plane.

2. Determine the location of the SP, and thereby the VPs and MPs by the 30° visual cone.

3. Transfer these dimensions to your drawing surface at an appropriate scale.

4. Find your starting point on the HML by measuring the distance it lies from the central vision line.

5. Using the HML and MPs, construct the perspective plan of the wedge take-up block.

6. Place the GL in relation to the HL to reveal the desired amount of the block's top.

7. Using the corner that touches the PP in the perspective plan, build the wedge take-up block in perspective.

Figure 6–10. A typical problem.

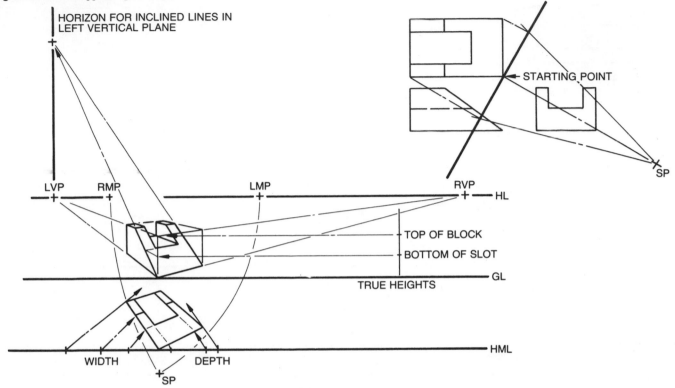

8. If the two parallel inclined edges were to be extended, they would meet at a vanishing point on a vertical line directly above the LVP. This vertical line is the horizon line for all lines lying in planes parallel to the left side of the object. This vanishing point can be used to assure that the edges of the slot that should be parallel to the outside of the plane are truly parallel.

Intersection of Plane Surfaces

Earlier I mentioned that if you could locate a point on, behind, or in front of the picture plane, you could draw any perspective. This is true; but with another set of conditions—namely, what happens to planes in perspective—you can really become proficient in perspective drawing.

The intersection of any two plane surfaces will be directed to the intersection of the two plane's horizon lines.

In Figure 6-11, two inclined planes have been established on top of an object. If the horizon lines for these planes were constructed and their intersection found, the intersection of the two planes could be directed there.

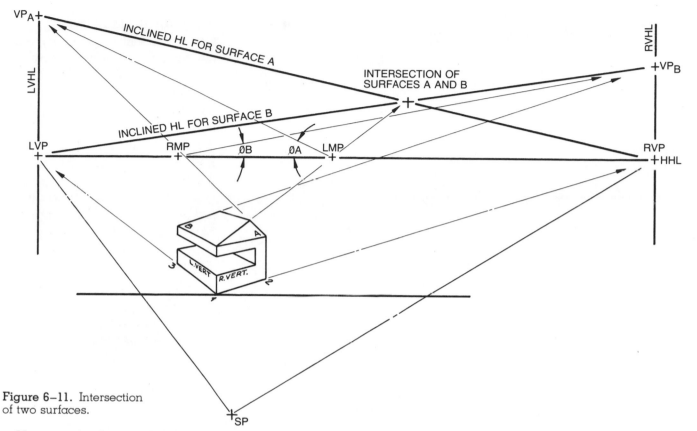

Figure 6–11. Intersection of two surfaces.

clined edges of surface A lie in left vertical planes, a line from the LMP to their vanishing point describes their normal inclination. Likewise, since the angled edges of surface B lie in right vertical planes, a line from the RMP to their vanishing point describes the angle surface B makes with the horizontal.

Perspective with the Ease of Isometric

With an understanding of the relationship of observer, object, and picture plane, and with the ability to simplify fundamental projection into construction aids, your knowledge of further simplification can lead you to a method of perspective almost as simple as isometric drawing.

This method is based on measuring point perspective using equal scales for measurement along the vertical and base line horizontals at the picture plane. Separating vanishing points by a comfortably large distance (often dictated by the size of the drawing surface) pulls the observer away from the picture plane. This broadens the area of acceptable distortion and minimizes the need to actually establish the PP–SP distance.

How might this work? Take the intersection of the right vertical plane and the ground as an example. Theory tells us that the line of intersection between the two, line 1-2, will be directed to the intersection of the horizon line for the ground and that of the right vertical plane. The horizon line for the ground is the HHL. The horizon line for the right vertical plane is the RVHL and runs vertically through the right vanishing point. The inter-

section of the two is the RVP, and, indeed, it is the point to which line 1-2 is directed.

All lines in the right vertical plane vanish somewhere along the RVHL. All lines in the left vertical plane vanish somewhere along the LVHL. The inclined edges of surfaces A and B are examples of these lines. If you were to extend the inclined edges of the planes as shown, their vanishing points on the vertical horizon lines could be found. By extending the hori-

zontal boundary of each plane to the HHL as shown and by finding two vanishing points of lines on planes A and B, you have found their respective horizon lines. Where these two horizon lines converge is the point to which the line of intersection is directed.

Also shown in Figure 6-11 is the rotational method for determining the true inclination of angled surfaces. Since the in-

117

A DIMETRIC-LIKE SETUP

Figure 6-12 presents a perspective situation where two of the faces of a cube will be equally exposed. To set up a perspective:

1. Establish an HL that separates the RVP and LVP by a large but comfortable distance.
2. Bisect this distance either physically or at scale.
3. Locate the SP on an arc through the VPs.
4. Determine the RMP and LMP in the conventional way.
5. Establish the HML in relation to the HL to yield the desired object view.
6. Pick a scale that results in a drawing of the desired size. To keep distortion within limits, avoid drawing outside the measuring points.

Figure 6-13 demonstrates several alternate views achieved by changing the HML–HL distance.

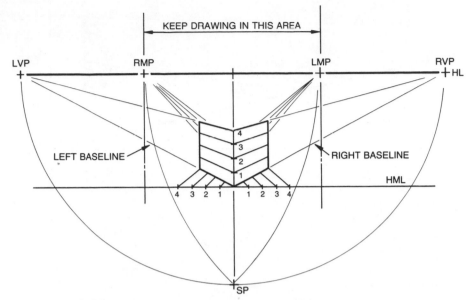

Figure 6–12. Perspective with two equally exposed faces.

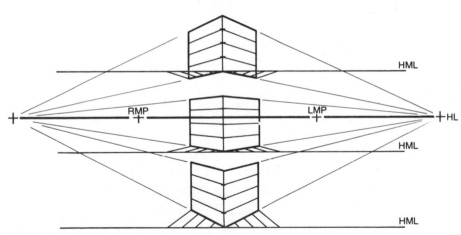

Figure 6–13. Alternate views.

A TRIMETRIC-LIKE SETUP

More interesting perspectives are formed when faces are not equally exposed. A common setup is to incline the object to the picture plane at 30° and 60°, as was done in Figures 6-7 and 6-9. In this situation the VPs and MPs conform approximately to the relationship shown in Figure 6-14. This allows the illustrator to set up an effective perspective without noting the position of the SP. To set up the 30°–60° perspective:

1. Separate the VPs by a comfortably large distance.
2. Divide this distance by half, a quarter, and an eighth to yield a short face on the side of the eighth division.
3. Establish an HML at an appropriate distance to the HL to yield the desired object view (Figure 6-15).
4. Pick a scale that results in a drawing of desired size. To keep distortion within limits, avoid drawing outside the measuring points.

Various upward or downward attitudes can be established by changing the relationship of the HML and HL, as shown in Figure 6-16.

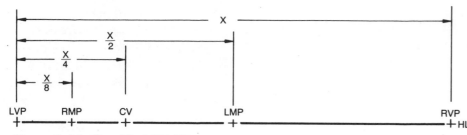

Figure 6–14. Simplified 30°–60° setup.

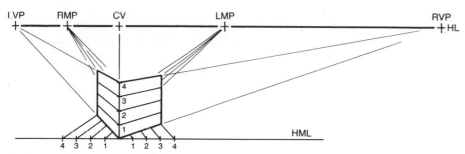

Figure 6–15. 30°–60° perspective construction.

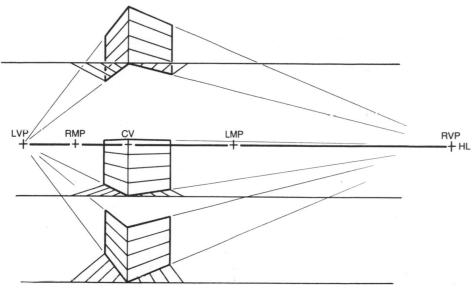

Figure 6–16. 30°–60° alternate views.

These situations (equal exposure and 30°–60° left and right) will cover most general perspective applications. For your use, several have been reproduced in Appendix D. The distances between VPs and MPs can be increased with your dividers to an appropriate scale for your drawing surface.

With your knowledge of setting up a perspective, of points and how they can be found, and of planes and how they intersect, you should now have the confidence to do perspective illustration.

Perspective is a specialized pictorial image. It requires somewhat different skills than axonometric drawing. The fact that you cannot easily measure directly on the perspective prohibits widespread use of perspective in technical situations. There are times, however, that only a perspective image can accomplish the goals of the drawing. These are the times that you must be prepared to construct a perspective drawing.

List of Terms

Because many perspective terms only appear in their shorthand form, knowledge of the terms and what they mean is very important. The terms used in this chapter on perspective drawing include:

angular perspective
base line
cone of vision
fundamental projection
ground line
horizon line

horizontal measuring line
measuring line
measuring point
measing point perspective
measuring wall
oblique perspective

parallel perspective
picture plane
station point
vanishing point
visual ray
visual ray perspective

Problems for Further Study

The perspective problems in this section test your ability to make perspective decisions and work the mechanics. The first problem requires no decisions on your part and is set up ready for you to work using the visual ray method. The second problem is also visual ray but requires you to make decisions as to place-ment of the SP, GL, and HL. Problem three is a measuring point perspective, again already set up for you to work. Problem four requires that you start with a clean sheet of paper, making decisions and doing the mechanics on your own.

1. Using the visual ray method, complete the perspective view of a bearing retainer as shown. Make careful note of where the object touches the PP.

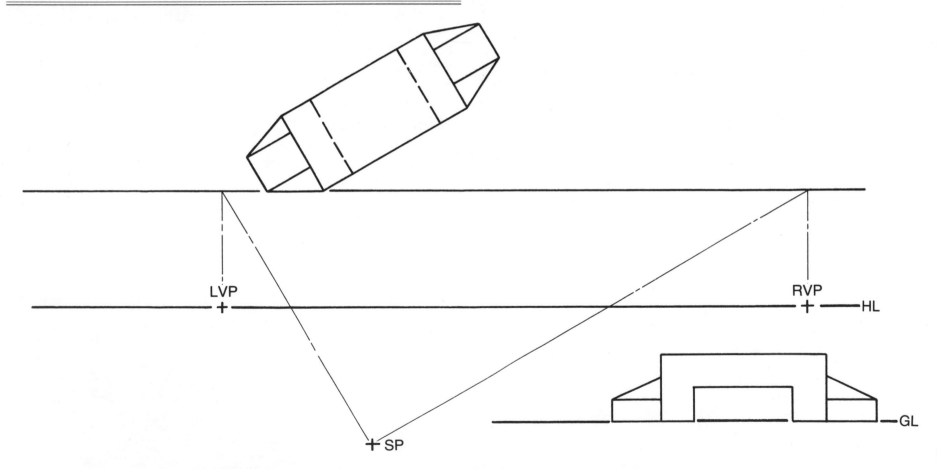

2. Given these detailed views of a U-hanger, complete a visual ray perspective that will show the object advantageously. Select an appropriate scale, orient the hanger to the PP, and place the HL and GL to yield a pleasing view of the object.

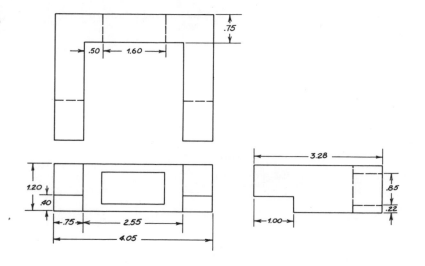

3. The necessary information has been presented to complete a measuring point perspective of a corner weldment. Note that the distance from the starting point to the central vision line has been established, with the right and left base lines shown. Complete the perspective plan first. Then build the object, starting with the front corner as true height.

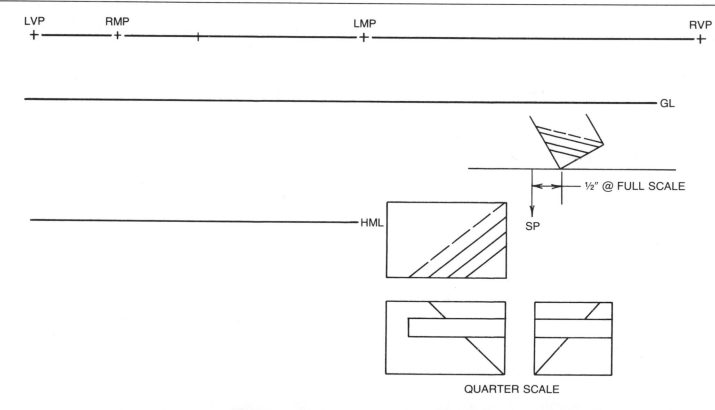

QUARTER SCALE

121

4. Given these detail drawings of a T-link at twice scale, complete a measuring point perspective at an appropriate scale. Decide on object orientation and produce a scale sketch to find the SP. Establish the HL, MPs, GL, and HML. Complete a perspective plan and build the object keeping in mind where *you* choose your starting point.

5. Given these detail drawings, construct perspective views that show the objects to their greatest advantage.

(a)

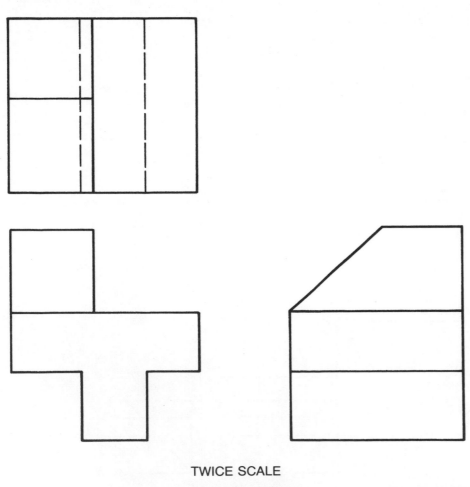

TWICE SCALE

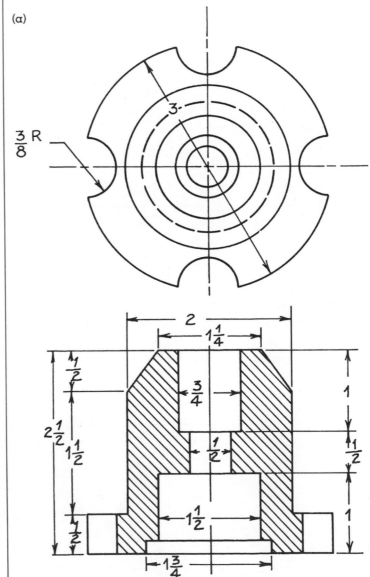

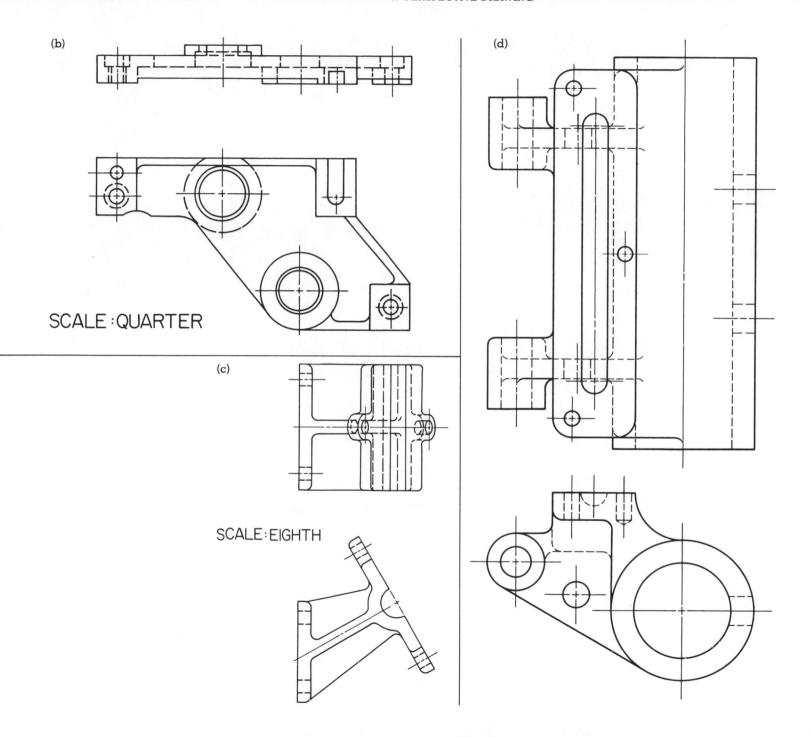

(b)

SCALE : QUARTER

(c)

SCALE : EIGHTH

(d)

CHAPTER SEVEN Presentation of Illustrations

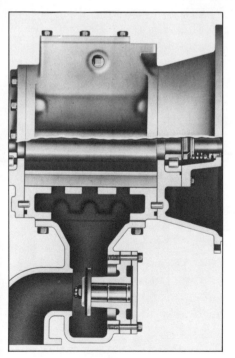
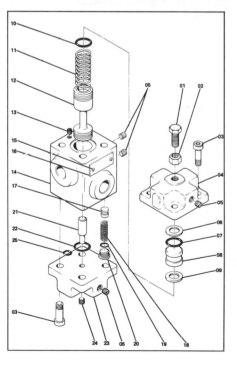
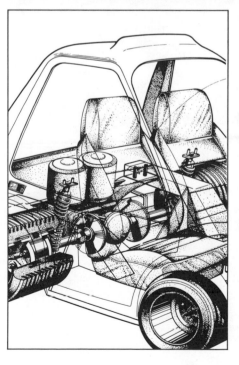

There was a time when illustrators were separated on the job into layout illustrators, inking illustrators, and airbrush illustrators. This was an efficient way to produce a volume of illustration, but it kept the artists from developing a full range of illustration skills: the ability to work from drawings, photographs, sketches, and life. Accurate construction, effective line rendering, and full-tone treatment should be at the command of each technical illustrator.

Even though it is best not to become overspecialized in one area of illustration, it is natural to become better in one area than the others. There is no aspect of illustration that makes a bigger impression than the way the drawing is rendered—*if* it is based on an accurate underdrawing. Rendering adds snap and personality to the illustration, bringing some parts forward, allowing others to recede. An illustrator's presentation style is unique to a point—it must still reflect the technical attributes of the subject.

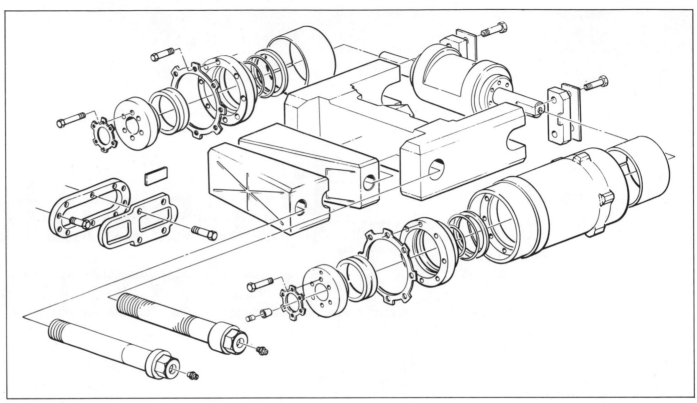

Figure 7–1. Effective line rendering.

The Psychology of Presentation

Technical illustrators make art for other people. In doing so, they set themselves apart from the kinds of artists who make art for themselves. Working for others, they need to understand many things about the client, the client's product, and the way the illustration will be used.

WHO IS THE CLIENT?

Clients differ in their products, organization, and size. Many industries have traditional uses for illustration as well as established methods of drawing. Clients also have expectations built on previous illustration experience. These traditions and expectations influence what the finished illustration will look like, how much it will cost, and how fast it must be turned around.

WHAT IS THE PURPOSE?

An illustration to be used in an informal meeting will obviously be different from one used to acquire funding for a multi-million-dollar power plant. One illustration may be used for parts identification, while another may be used to guide assembly. Thus, the purpose of the illustration often determines the nature of the presentation.

STORAGE, TRANSMISSION, AND REPRODUCTION

It would be a shame to do a continuous tone airbrush illustration only to find that it was to be photocopied, or screened at 60 lines to the inch and then microfilmed. Knowing the intended method of reproduction is important in both line and continuous tone illustration. It is also important in choosing the techniques of line illustration to use. What will the final reproduction size be? Will the detail and fine line work hold up, or will it be lost and block up?

Industry uses a broad range of methods and equipment to store, transmit, and reproduce drawings—both engineering drawings and technical illustrations. These differing methods may be found in almost any combination in the same industry group. From the less sophisticated to the most current, they are:

1. *Hand copying*. A time-consuming and expensive practice, hand drawing is still the only way to copy damaged originals or drawings with linework too weak to copy by any other method.

Figure 7–2. Methods of storing and transmitting drawings.

2. *Blueprints*. Producing white lines on a blue background, blue printing uses chemicals and a light source to make copies.
3. *Blueline prints*. Blueline prints are produced by a vapor process and are easier to read and more economical than blueprints.
4. *Electrostatic prints*. Black lines on paper or vellum are produced by dry copying

methods. If the carbon dust is properly fixed, prints are nearly permanent.
5. *Photographic negative*. Prints are made directly from the negative which is usually mounted in a computer aperture card; the original is kept in a historical file.
6. *Electronic storage*. Drawings can be stored in the form of a magnetic code and accessed

through electronic graphics devices, including CRT screens, plotters, and printers. Drawings can be sent across phone lines or by microwave transmission.

Companies usually update their drawing equipment only when an economic benefit can be demonstrated. They are in business to make money, and if updating their equipment can be shown to increase their profit, they are likely to consider the change. Thus, two similar companies, even ones across the street from each other, may have widely varying methods for storing, retrieving, and transmitting their drawings.

HOW MUCH TIME IS AVAILABLE?

Line is usually faster than continuous tone illustration; pencil is faster than ink. The amount of time (called *lead time* when it includes the time before you actually get the assignment) available determines to some extent the choice of presentation technique. The time that remains after doing the construction drawing is most critical. A particularly difficult drawing may mean a change in presentation technique if most of the time available must be used in figuring out the detail drawings and in doing the construction, leaving no time for airbrush or other time-consuming techniques.

HOW MUCH MONEY IS COMMITTED?

Money means board time and support services such as photography, typesetting, and subcontract services. Technical

illustration is *labor intensive;* 90% or more of the cost of an illustration is wages. Most illustrators can look at a job and estimate both its dollar value and the time needed to complete it. The Society for Technical Communications (STC) publishes *Estimating Illustration Costs—A Guide,* a handy reference for estimating time. No mention is made of the dollar per hour rate, so the guidelines are applicable anytime, anywhere. A page from the STC guide is reproduced in Figure 7-3.

HOW MUCH SKILL IS AVAILABLE?

Each illustrator should know his or her limitations. Some work may be subcontracted to others who have abilities the illustrator does not. With limitations and options in mind, the illustrator must consider the factors of time, money, and client needs when choosing the presentation technique. He or she should of course try new techniques but should not use them on jobs until the techniques have been perfected—certainly not on the first job for a client.

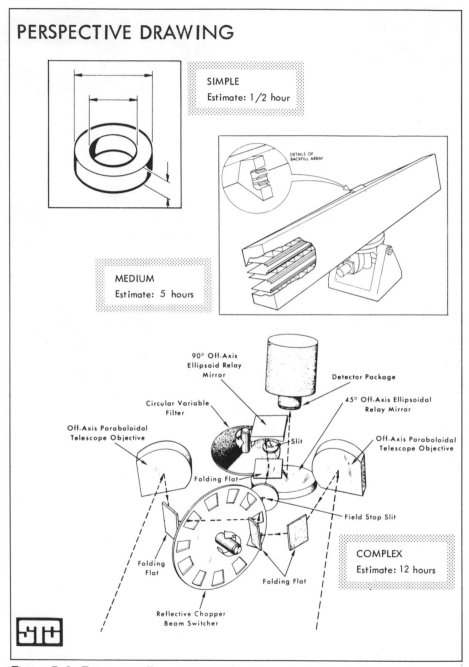

Figure 7–3. Estimating illustration cost by time required to do the drawing.

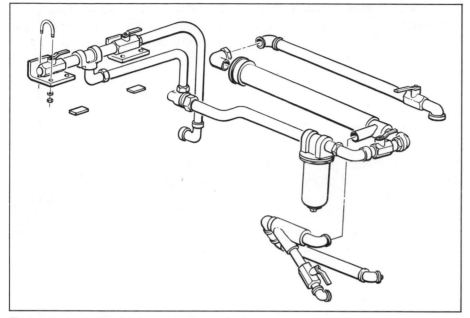

Figure 7–4. Appropriate rendering technique.

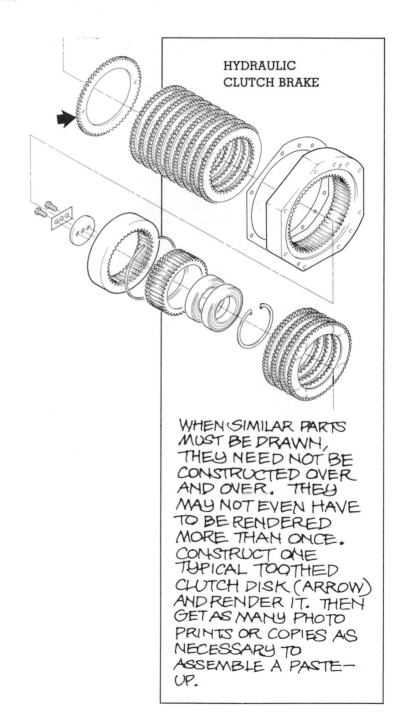

HYDRAULIC CLUTCH BRAKE

WHEN SIMILAR PARTS MUST BE DRAWN, THEY NEED NOT BE CONSTRUCTED OVER AND OVER. THEY MAY NOT EVEN HAVE TO BE RENDERED MORE THAN ONCE. CONSTRUCT ONE TYPICAL TOOTHED CLUTCH DISK (ARROW) AND RENDER IT. THEN GET AS MANY PHOTO PRINTS OR COPIES AS NECESSARY TO ASSEMBLE A PASTE-UP.

What Makes an Effective Presentation?

With these six factors in mind, you can see what makes a good presentation. At the center is an illustration that "works," which means that the purpose or intent of the illustration has been met. The characteristics of an illustration that works are:

1. *The drawing is accurate.* It accurately reflects the form, surface finish, and material.
2. *The drawing is clear.* Accuracy without clarity does not aid the illustration. Certain parts may be omitted, phantomed, or sectioned for the sake of clarity.

Some features may be drawn out of scale to make the drawing more clear.
3. *The composition is pleasing.* The design of the illustration—how it fills the space, the choice of viewing direction, how parts are arranged—contributes to how well it works.
4. *The technique is appropriate.* The choice of line weight, highlights, shading, and tone is appropriate to the material, scale, mass, and surface.

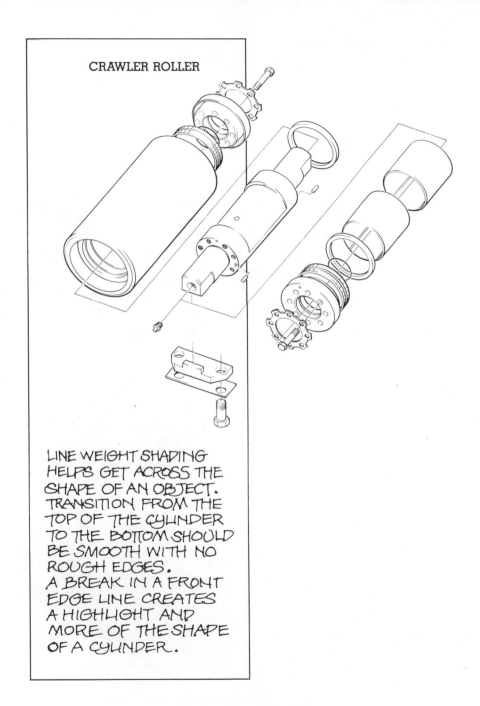

CRAWLER ROLLER

LINE WEIGHT SHADING HELPS GET ACROSS THE SHAPE OF AN OBJECT. TRANSITION FROM THE TOP OF THE CYLINDER TO THE BOTTOM SHOULD BE SMOOTH WITH NO ROUGH EDGES. A BREAK IN A FRONT EDGE LINE CREATES A HIGHLIGHT AND MORE OF THE SHAPE OF A CYLINDER.

Choosing the Drawing Format

Final reproduction size influences many decisions that an illustrator makes concerning type of illustration (isometric, dimetric, trimetric, or perspective), media (pencil, ink, paper, line or continuous tone), and individual line weights. There are many variables that must be considered when deciding final illustration size and the reduction required to get there. These variables fall into three main categories: policy, skill, and equipment.

POLICY

Variables beyond the control of the illustrator may include:

1. Company policy—to conform to existing documentation.
2. Graphic arts requirements—press size, plate size, available paper sizes, and so on.
3. Industry-wide conventions—some industries use foldouts exclusively, some not at all.
4. National conventions—metric and standard American drawing and paper sizes.

SKILL

All of these policy factors influence drawing and pen sizes. Beyond these, there are variables that deal with the personnel who will work on the illustration. These variables include the illustrator's own skill, the photographer's skill, and the printer's skill.

If you know the people who will work on the illustration, you can compensate for problems in any of these skill areas. If you are unfamiliar with the illustrator, photographer, or printer, you should secure examples of their work on jobs similar to the one they will be working on.

EQUIPMENT

The type of equipment available also influences how great a reduction may be made. Such equipment includes:

1. Illustration media—papers, film, inks, pencils, screens, tints, and so on. All have different attributes.
2. Photographic media—film speed, negative size, printing method, copy stand design, lighting, and so on. All contribute to accuracy and detail.
3. Printing equipment—plate-making materials and method, accuracy of the printing press, quality of the ink and paper, and so on. All contribute to drawing reproduction quality.

HOW REDUCTION AFFECTS PEN SIZE AND LINE WEIGHTS

Line weights on the final printed illustration must not "break up" or otherwise become faint or unreadable. You should consult Table 7-1 when choosing line weights for the original illustration. Of course, policy, skill, and equipment will temper your decisions.

The pen sizes in Table 7-1 are approximate and assume a new pen, constant pen size, constant pen speed, a perpendicular orientation to the drawing surface, and controlled ink, humidity, and temperature. The three pens listed are among the most popular.

To use the table, think of the size line you want on the final reproduction. If the final reduction is 50% of the original, a number 2 pen will reduce down to a 2×0 line (.10mm), and so forth. The missing values reflect lines thinner than 5×0 and should be avoided for most reductions.

Table 7-2 provides data on reductions from standard drawing sizes. Let's say that your final illustration will be printed on 8½ × 11 paper, filling a 6 × 9 format. What are your options within the percentages given? From Table 7-2:

Table 7-1. Common Pen Sizes and How They Are Affected by Reduction

Castel TG		Pens Mars 700		Rapidograph 3065		Percent of Original Size					
#	mm	#	mm	#	mm	86	77	66	50	33	25
3×0	0.10	5×0	.13	5×0	.134	6×0	—	—	—	—	—
2×0	0.20	4×0	.18	4×0	.19	5×0	5×0	—	—	—	—
—	—	3×0	.25	3×0	.25	4×0	4×0	5×0	—	—	—
0	0.30	2×0	.30	2×0	.30	3×0	3×0	4×0	5×0	—	—
—	—	0	.35	0	.35	2×0	2×0	3×0	4×0	—	—
1	0.40	1	.45	1	.46	0	0	2×0	3×0	5×0	—
2	0.50	2	.50	2	.56	1	1	0	2×0	4×0	5×0
3	0.80	3	.80	3	.81	2	2	1	0	3×0	4×0
4	1.00	4	1.20	4	1.14	3	3	2	2	0	2×0
7	2.00	6	2.00	6	1.70	4	4	4	3	2	1
—	—	—	—	7	2.00	6	6	4	4	2	1

Table 7-2. Standard Drawing Sizes and Reductions

Drawing Size		Percent of Original Size					
		86	77	66	50	33	25
A	9 × 12	7.74 × 10.32	6.93 × 9.2	5.94 × 7.9	4.5 × 6.0	2.94 × 3.9	2.25 × 3.0
B	12 × 18	10.32 × 15.48	9.24 × 13.86	7.92 × 11.88	6.0 × 9.0	3.96 × 5.9	3.00 × 4.5
C	18 × 24	15.48 × 20.64	13.86 × 18.48	11.88 × 15.84	9.0 × 12.0	5.94 × 7.9	4.50 × 6.0
D	24 × 36	20.64 × 30.96	18.48 × 27.72	15.84 × 23.76	12.0 × 18.0	7.92 × 11.88	6.00 × 9.0
E	36 × 48	30.96 × 41.28	27.72 × 36.96	23.76 × 31.68	18.0 × 24.0	11.88 × 15.84	9.00 × 12.0
F	28 × 40	24.08 × 34.40	21.56 × 30.80	18.48 × 26.40	14.0 × 20.0	9.24 × 12.00	7.00 × 10.0

Size	Reduction	Final Size
9 × 12	.77	6.93 × 9.24
12 × 18	.50	6.0 × 9.0
18 × 24	.33	5.94 × 7.92
24 × 36	.25	6.0 × 9.0

A 12 × 18 drawing (50% reduction to achieve final size) will give you an easy proportional reduction and an original size that is not so large as to be unmanageable.

To determine line weights for your 12 × 18 drawing, look at Table 7-1 under the 50% column. The table shows that to get a 5×0 line on the final reduction, you should use a 2×0 line on the original. Using a pen smaller than 2×0 will result in a line thinner than 5×0 on the reduction, which may not reproduce. The exception is with the Castel TG pens. Note in Table 7-1 that the 3×0 Castel pen is slightly thinner than the Mars or Rapidograph 5×0.

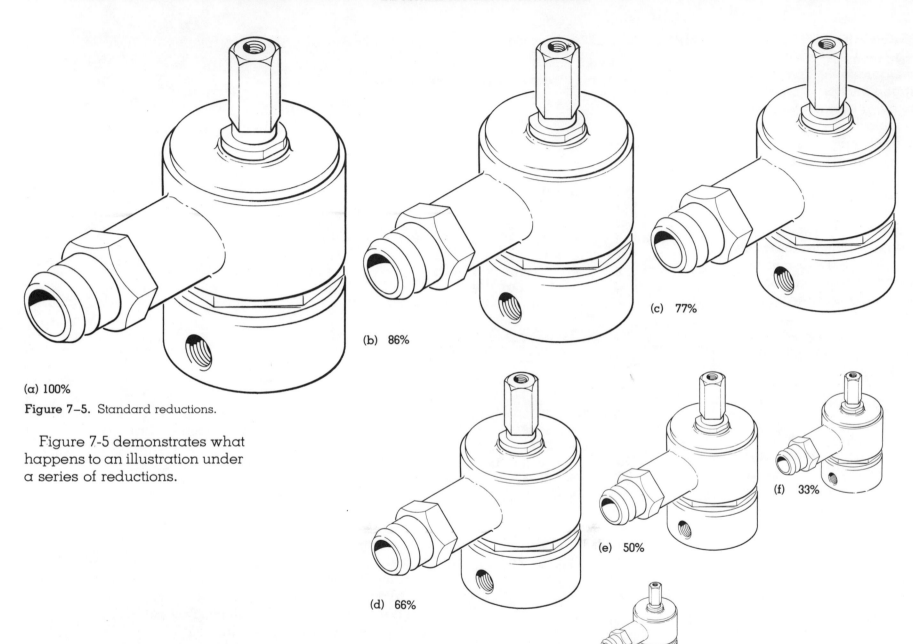

(a) 100%

Figure 7–5. Standard reductions.

Figure 7-5 demonstrates what happens to an illustration under a series of reductions.

(b) 86%

(c) 77%

(d) 66%

(e) 50%

(f) 33%

(g) 25%

Presentation Techniques

Presentation techniques include line rendering, block shading, cross hatching, stippling, and continuous tone.

LINE RENDERING

The most common rendering technique is called *line rendering*. In it, the lines that define edges or boundaries of the form are given special treatment according to a line weight strategy. This strategy is illustrated in Figure 7-6, in which all objects are rendered consistently. With such a strategy, drawings done at different times can be scaled and combined without lengthy adjustment of the linework.

The sun in Figure 7-6 has been established as coming down from the right, but it can be brought in from either the left or the right. The heaviest lines will be those on the bottom of the object away from the sun. These lines give solidity and define the limits of the form. I have assigned the value 3 to these lines, corresponding roughly to a number 3 technical pen. This value must be adjusted up or down depending on the scale of the drawing.

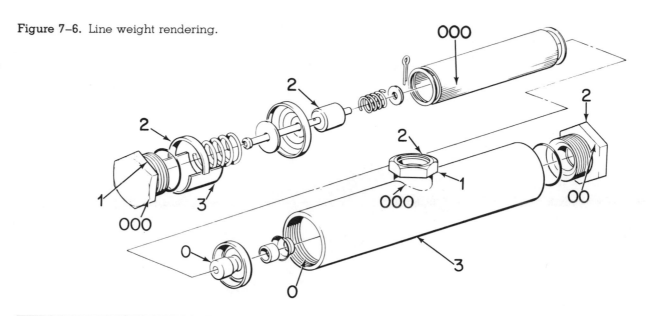

Figure 7–6. Line weight rendering.

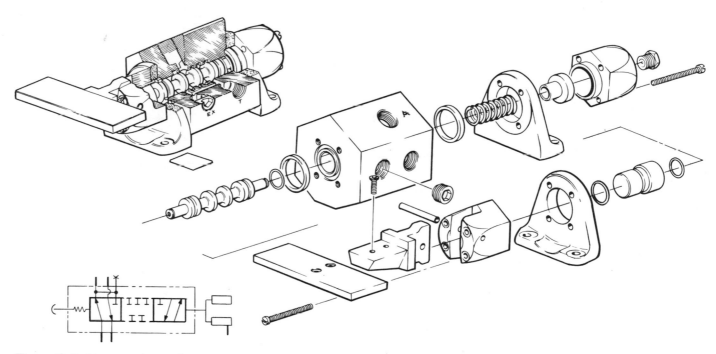

Figure 7–7. Line weight rendering.

133

Smaller, more delicate objects require thinner, more delicate lines.

The next thinnest line (2) defines the outside of the object but not on the bottom. A number 2 line separates the object from the background.

A number 1 line defines one part of the object as it overlaps another part of the object. If on the bottom, this line will be slightly heavier.

Lines 0–000 show planes as they intersect. If two planes were to touch making an angle of 180°, there would be no line formed. Lines 0–00 take care of situations in which planes come together making angles of 90°–179°.

The thinnest lines, broken at times to look even thinner, are reserved for defining curved surfaces. Figure 7-8 is an example of this line-rendering technique.

Although the strategy I have just described is not the only one for line rendering, it is one that has worked well for both students and professionals and is easy to keep in mind.

When rendering a drawing, it is difficult to know just where to start. Much of line weight technique involves applying

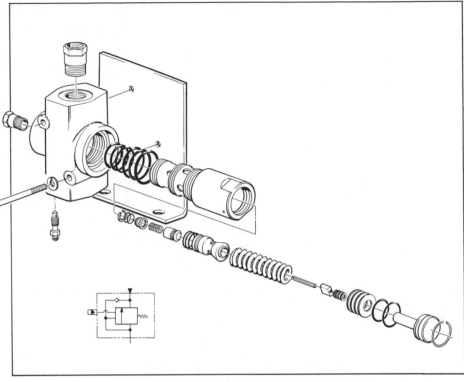

Figure 7–8. Line weight rendering.

line weight strategy to geometric form. But more than this, lines need to smoothly move from one thickness to another. In pencil, sharpened lead renders edges or 000 lines; as the pencil wears down, fatter lines are produced. When the pencil is well worn, outside lines can be done. In all cases—thick or thin—a 2H or H pencil produces a black line.

Starting at the wrong place or with the wrong technical pen or pencil point may force you to

compromise the line weight strategy. Keep these points in mind when line rendering in pen or pencil:

1. Base the rendering on an accurate under-drawing.
2. Pick a position for the source of light.
3. Start near, then go far.
4. Start thin, then go thick.

Figure 7-9 illustrates how this strategy can be used on a cylindrical pin. First, construct an accurate layout of the part. Next, decide where the sun is coming from. Start on the end nearest to you by retracing the outer ellipse, leaving a highlight at one o'clock. This brings the light source over your right shoulder. Move the ellipse guide down and to the right along the major axis, feathering the line up to the right and then the left. Make this a *smooth* transition—don't let the blending line overhang the first line. Sit back and look at this treatment. If the bottom isn't thick enough, repeat the blending process.

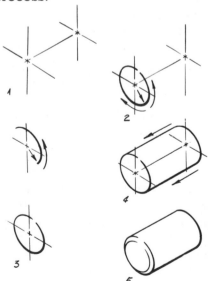

Figure 7–9. Steps in rendering a pin.

Do the same on the rear of the pin, keeping in mind that half of the ellipse will be hidden. When this is done correctly, a tangent element at the top will join lines of the same thickness; an element at the bottom will join a pair of thicker lines.

With ink, the same pen should be used for the thick and thin lines. Only in this way can you assure a smooth transition, which is characteristic of a well-rendered illustration. Through practice, you can attain consistent line weights.

After finishing the line weight rendering, you can use block shading if a stronger presentation is desired.

BLOCK SHADING

Block shading takes line weight one step further in defining the solidity of the object. It is very effective on cylindrical parts but is less so on prisms or other large forms, where it is easy to end up with an object that looks like it was painted black.

To avoid this problem, the minimum amount of block shading should be used. Figure 7-10 shows how the conjugate axes can be used to limit the area of the block shading. On more difficult objects, several rough tracings may be made to

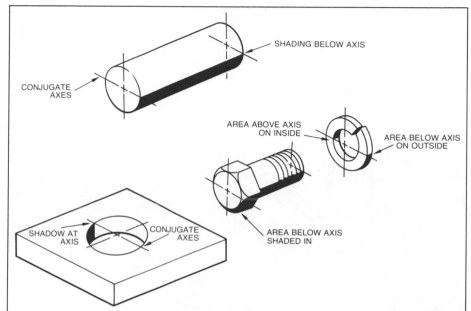

Figure 7–10. Conjugate axis shading.

try out possible shading before actually rendering the object.

On occasion, shading may have to be applied on a photograph without the benefit of having the construction axes to go by. In such cases, as in Figure 7-11, knowledge of the conjugate axes can keep you on the right track even when you don't physically locate them.

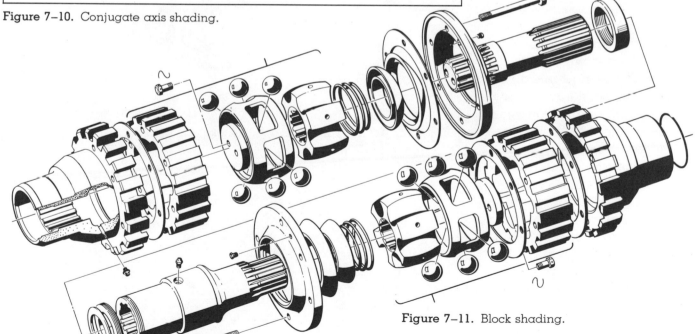

Figure 7–11. Block shading.

135

ILLUSION OF CONTINUOUS TONE

Rendering surfaces so that they appear as they would naturally can be done without actually making the tones continuous. This illusion of continuous tone allows the illustration to be reproduced as line copy without screening. Furthermore it saves time and money, since a full-tone illustration can be anywhere from two to five times more expensive.

Cross hatch rendering builds tone by placing surface lines close together for dark tones and further apart for lighter tones. The lines must be chosen with care so that the surface is defined correctly. It takes time, but the effect is well worth it. Cross hatch is most effective on plane or single-curved surfaces (cylinders and cones)—especially those that have been machined.

Stipple rendering (Figure 7-12) can be used to render more complex surfaces where block shading or cross hatching would be inappropriate. Stipple is more effective for rough, cast, or rubber surfaces. The dot pattern can be put down from a transfer sheet, but for ultimate control of the tone pattern each

dot should be done by hand with a technical pen. You can avoid building patterns of swirls into the stipple by working back and forth across an area. Or you can use the "cross-eyed technique," in which you let your eyes relax just enough to blur the image allowing you to spot open or closed areas.

CONTINUOUS TONE

No more than 15% of technical illustration is continuous tone, yet it is one of the most impressive techniques when done effectively. It is the subject for an entire course of study. An accomplished illustrator needs a great deal of practice in handling an airbrush, brush, or pencil; in mixing paints, dyes, or pastels; and in masking and frisketing. Figure 7-13 shows that results just as effective as airbrush can be attained using a brush and the right amount of skill. A fully airbrushed illustra-

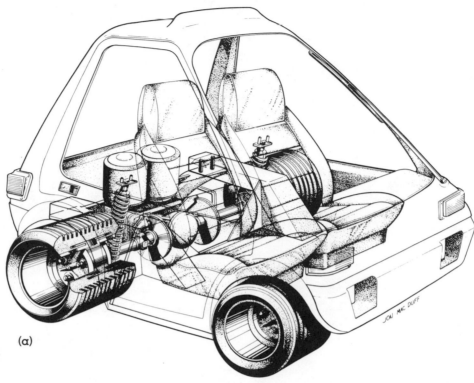

(a)

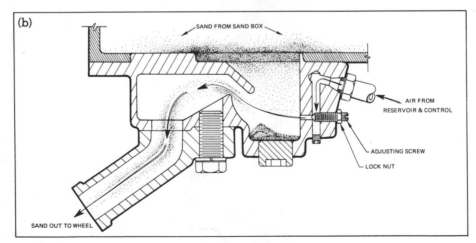

(b)

SAND FROM SAND BOX

AIR FROM RESERVOIR & CONTROL

ADJUSTING SCREW

LOCK NUT

SAND OUT TO WHEEL

Figure 7–12. (a) Stipple rendering. (b) Stipple technique.

Figure 7-13. Brush and pen technique. (Peter Trojan for *Popular Mechanics.* ©1976 The Hearst Corporation.)

AUGER GEAR CASE

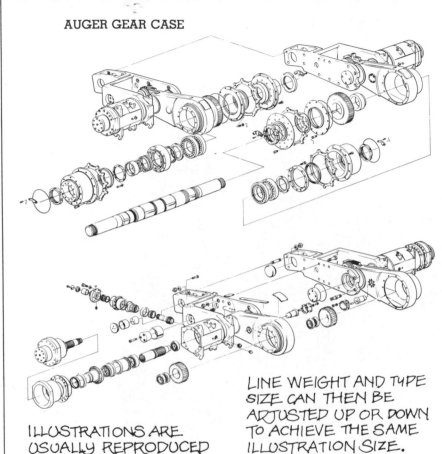

ILLUSTRATIONS ARE USUALLY REPRODUCED AT SPECIFIC REPRODUCTION RATIOS TO KEEP LINE WEIGHTS CONSISTENT FROM ONE DRAWING TO ANOTHER. MANY COMPANIES HAVE THEIR ILLUSTRATORS WORK NOT TO A SPECIFIC SIZE BUT TO ONE OF SEVERAL FORMAT RATIOS.

LINE WEIGHT AND TYPE SIZE CAN THEN BE ADJUSTED UP OR DOWN TO ACHIEVE THE SAME ILLUSTRATION SIZE. ON COMPLICATED ASSEMBLIES LIKE THIS ONE, THE FINAL DRAWING MAY BE A COMPOSITE REPRESENTING PIECES OF SEVERAL ILLUSTRATIONS, EACH AT A DIFFERENT SCALE. AFTER REDUCTION, THEY CAN BE ASSEMBLED TO MAKE A FINAL DRAWING.

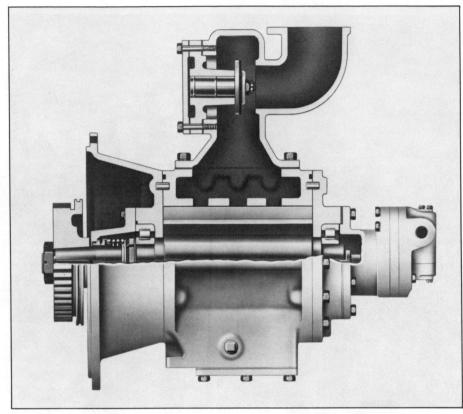

Figure 7–14. Airbrush technique. (Courtesy James Shough Visuals.)

tion, as in Figure 7-14, allows for slightly smoother transition of tone, especially over a large surface. Combining linework and airbrush can be particularly effective if a portion of the illustration is highlighted, as in Figure 7-15; the point of interest is highlighted in airbrush, making for a pleasing illustration.

In continuous tone illustration, values are spread from light to dark. The most common problem is bunching the values too closely together. When they are grouped at the light end, the rendering is pale and anemic. When they are grouped at the other end of the scale, the rendering is dark, with little detail. Thus, to be effective a continuous tone illustration must have values spread out along the scale from dark to light.

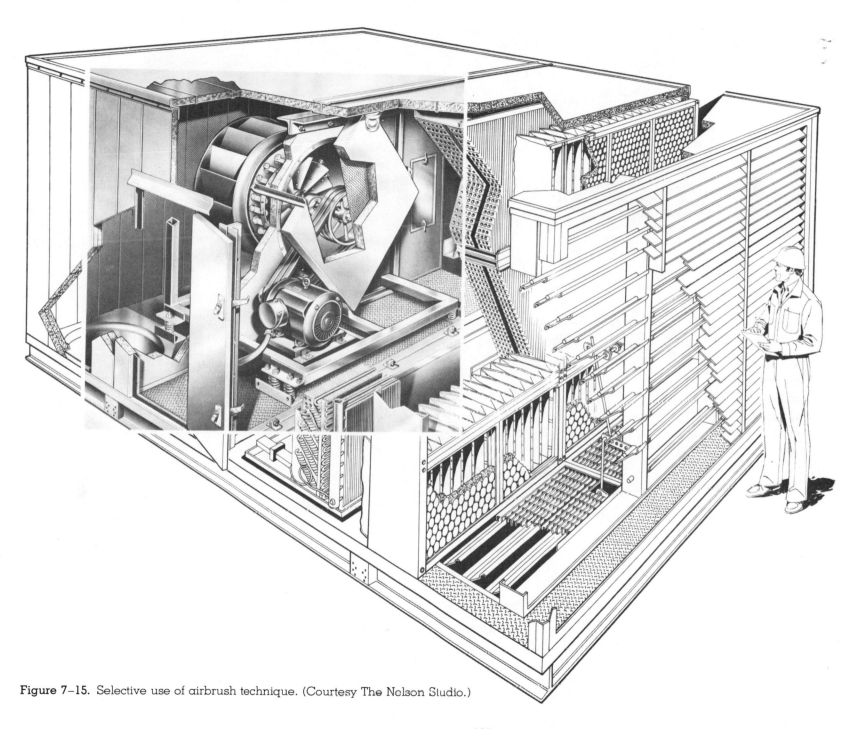

Figure 7–15. Selective use of airbrush technique. (Courtesy The Nelson Studio.)

(a)

(b)

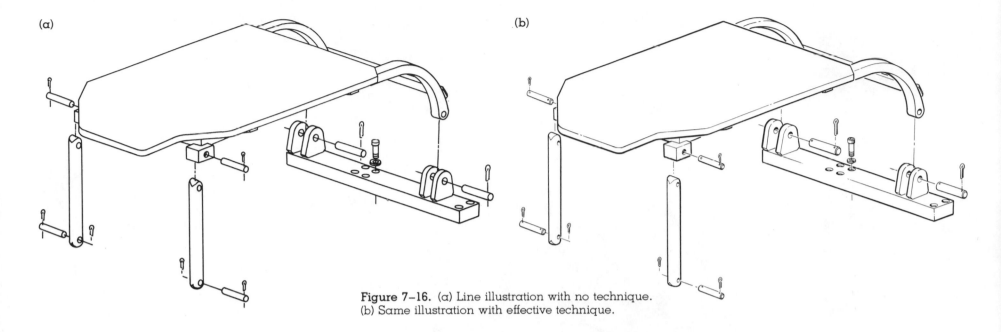

Figure 7–16. (a) Line illustration with no technique. (b) Same illustration with effective technique.

To quickly summarize, line illustration is the most important illustration technique, after learning accurate construction. You can see in Figure 7-16 the difference between a line illustration with no technique and one with effective technique. Not only does the effective drawing speak more closely of form and solidity, it is more easily reproduced. For most purposes, line illustration such as that in Figure 7-17 represents the most economical, rapid, and effective illustration technique. Master this technique first, then add cross hatch and

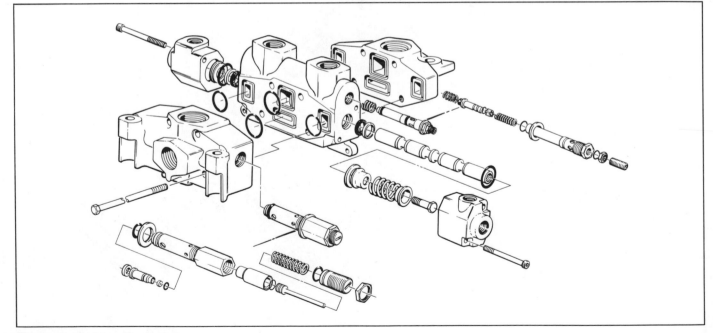

Figure 7–17. Line illustration.

stipple. If your interests direct you, and your talents lie in that direction, you may want to develop the skills to produce continuous tone illustration.

Design Illustration

When an industrial designer roughs out a sketch, trying to lend form to his or her ideas, lines flow. The sketch hardly ever looks like a finished illustration. This is as it should be. Designers best spend their time designing, not illustrating.

The design illustration speaks of spontaneity and freshness; it evokes a completely different response than a finished illustration. When desired, a finished illustration can be made to look like it came right off a designer's drawing board. The illustration shown in Figure 7-18 is meant to evoke feelings of a design illustration. You can see the features that mark a design illustration:

1. Centerlines and axes are dominant as the foundation or structure of the illustration.
2. Lines are incomplete or overlap.

3. There may be accompanying text or graphics to help explain the design.

REMOVABLE CAB

SIDE ENTRY

COUNTER WEIGHT

ARTICULATING FRAME

RONE

Figure 7–18. Design illustration. (Courtesy Frank Rowe.)

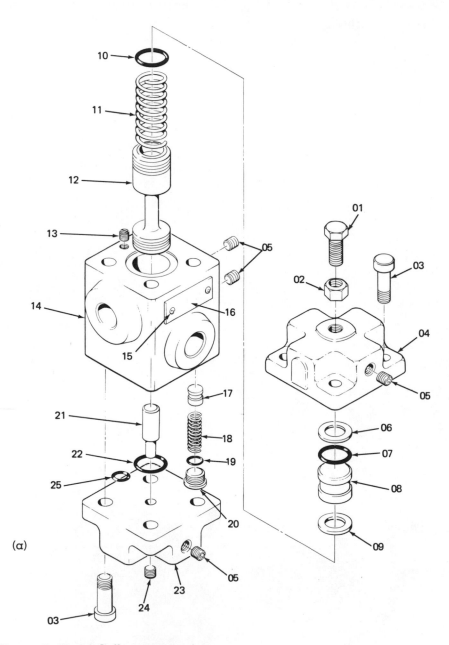

(a)

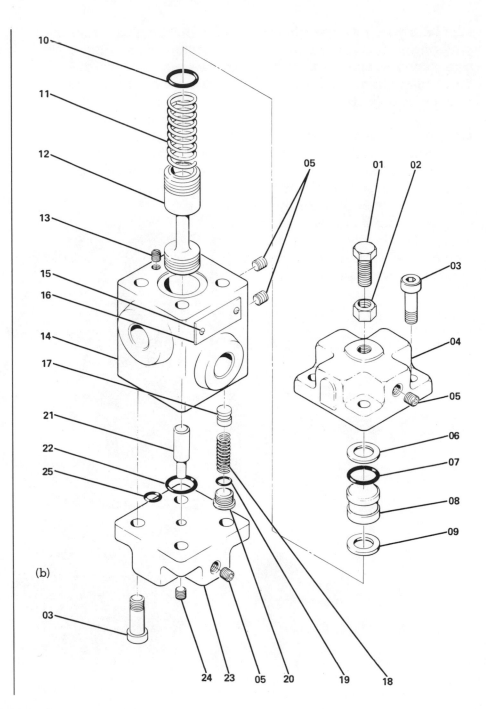

(b)

Figure 7–19. (a) Call-outs in random pattern. (b) Call-outs in linear pattern.

Call-Outs or Mark Numbers

Many illustrations are used for parts identification, maintenance, or assembly. In such illustrations individual items need to be identified and set apart from other items or included as being like similiar items. This is the only way that all of the parts in an assembly can be accounted for. The identification is usually done by numbers, one assigned to each line item, with leaders pointing to the items in question. These numbers can be enclosed in circles ("balloons") to set them off, a practice best suited to drawings that have few items to be identified—the balloons may take up too much room on more complicated drawings. Numbers as well as leaders with arrow heads can be found in various sizes on transfer or cutout sheets. They can be rapidly

applied, burnished, and even changed if need be.

The numbers may be arranged either randomly or in a pattern. Figure 7-19 shows the same illustration with the callouts first done in a random pattern and then in a linear pattern.

Leaders should be kept short yet without crowding the drawing. Long leaders can detract from the illustration. It is better not to have leaders cross one another. This can be avoided by rearranging mark numbers or, if necessary, parts of the drawing. Keep mark numbers outside the object unless absolutely necessary.

The placement of call-outs is often dictated by company or client policy. Such policy is established so that all illustrations are visually compatible and so that individual parts will be easier to identify from drawing to drawing.

List of Terms

block shading	cross hatch rendering	line rendering
blueline prints	design illustration	mark numbers
blueprinting	electronic storage	photographic negative
call-outs	electronic prints	stipple rendering
continuous tone	lead time	

Work Problems

Making industrial technical illustration consists of breaking complex systems or machines into manageable groups or individual items. The items can then be assembled back into the whole.

The problems in this appendix begin with single pieces and progress to more complex parts and assemblies. If you start with the single items first, you will gain the skill to attack one of the major pieces of illustration, such as that found in problem A-4.

Choose among the axonometric scales and grids in Appendix D or make your own using Table C-4 or C-5.

The best experience you can get is working with engineering drawings—full-sized ones like those you would find on the job.

The work problems in this appendix are just a start. You and your instructor are encouraged to ask local industry for additional problems after you have worked the problems in this book.

Engineering Drawings. In problems A-1 through A-6 detail drawings have been reproduced with accurate dimensions. You can work out a scale for a particular drawing by taking a known distance and expanding it into a scale. All distances can then be determined.

The drawings are arranged with bills of materials to identify individual parts. It is best to work from the assembly drawing and then move to the details. Spend some time with the drawings before you actually start your illustration. Make freehand sketches or color individual parts to help you visualize the assembly. Choose the orientation that best displays the object, then begin careful construction.

Photographs. For problems A-7 through A-9, use an overlay material (acetate, film, vellum) to render the equipment in each photograph. Omit unnecessary or confusing detail. Determine a center of focus and allow the drawing to fade or vignette. Use a line illustration technique or, if necessary, block shading.

Computer Underlays. Computer underlays can be produced in a fraction of the time required for the same drawing to be done by hand. If another view of the same situation is required, the drawing can be done in a matter of seconds. Of course, an illustrator must put the finishing touches on the underlay.

Problems A-10 and A-11 display several computer-generated illustrations for you to finish. Use a suitable overlay material, determine ellipse exposure and direction of the sun, and produce a line rendering based on the computer-generated underlay.

Note that the computer approximates ellipses in a series of short steps. Straight lines are plotted most accurately. Whether or not a line ends at the correct spot depends on the accuracy of the instructions or directions given to the computer and on the accuracy in which the geometry of the object has been defined.

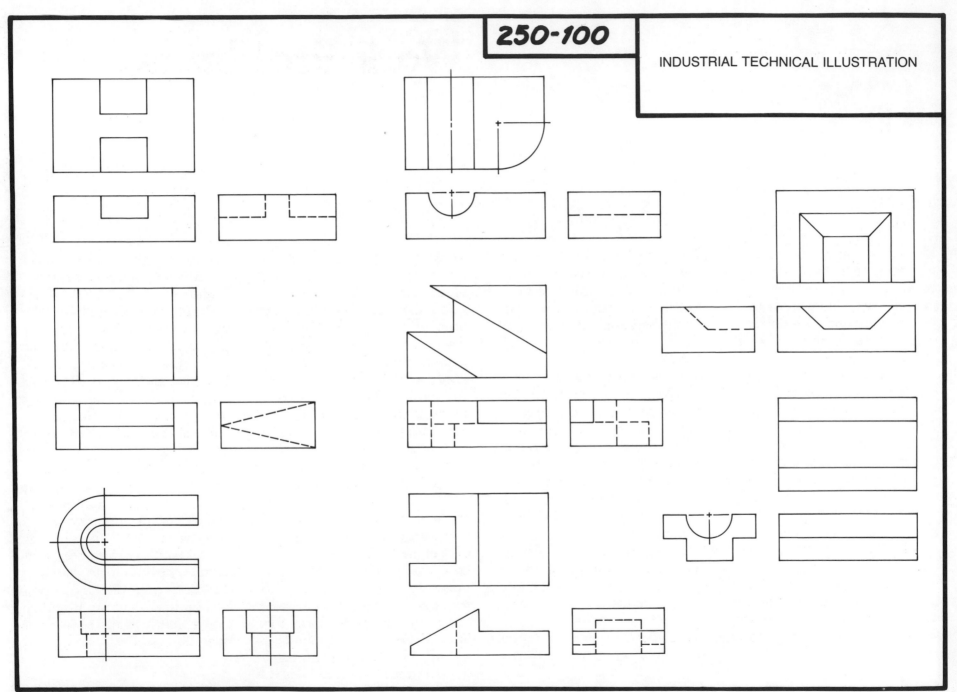

250-100

INDUSTRIAL TECHNICAL ILLUSTRATION

Problem A-1. Basic orthographic objects.

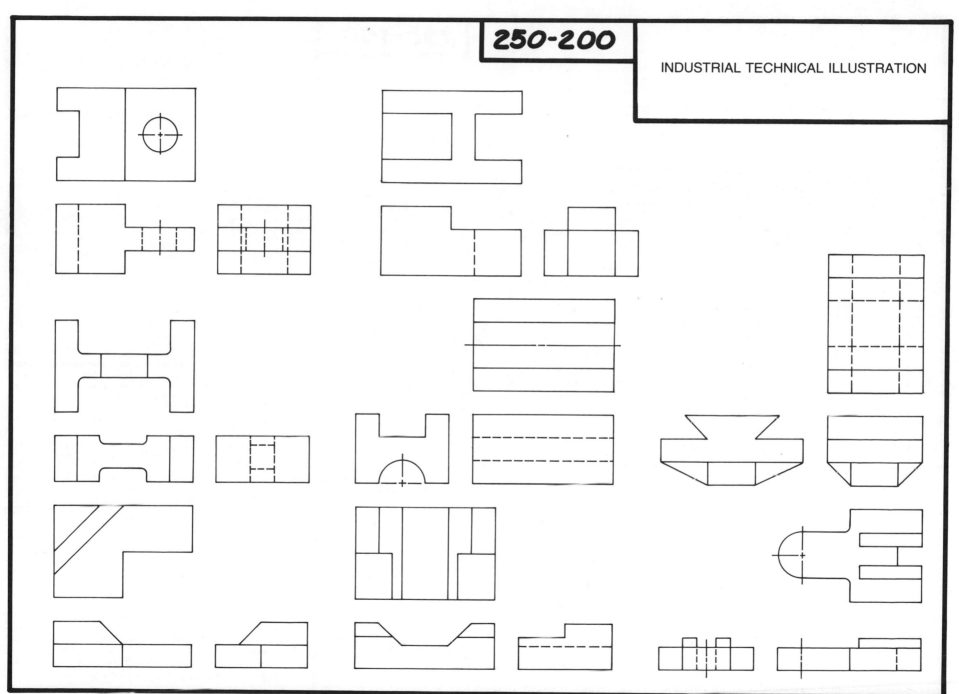

Problem A-2. Basic orthographic objects.

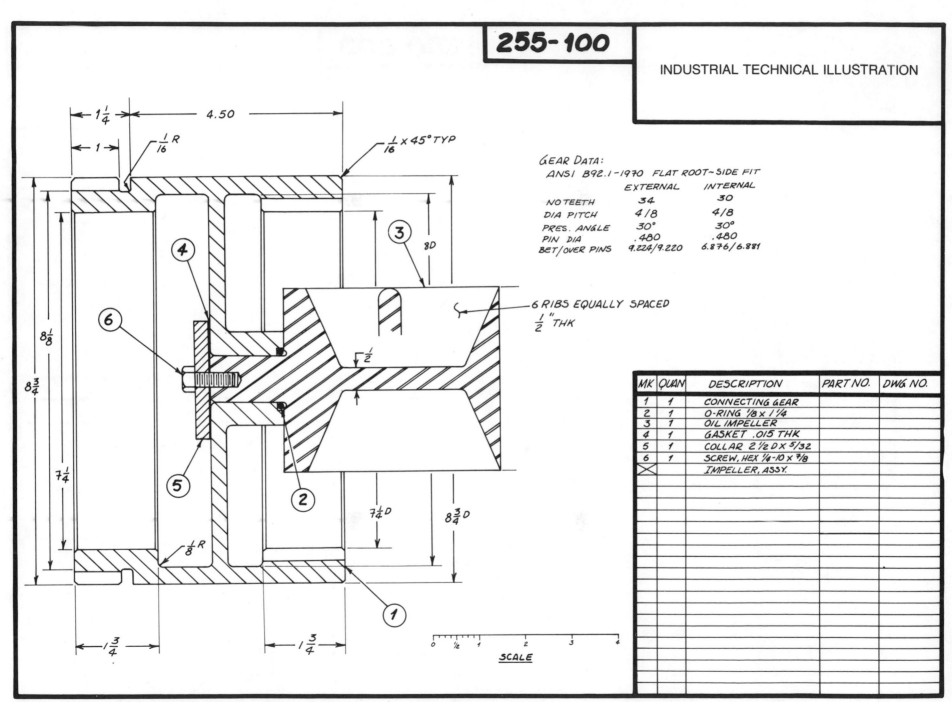

255-100

INDUSTRIAL TECHNICAL ILLUSTRATION

$1\frac{1}{4}$ — 4.50
1
$\frac{1}{16}R$
$\frac{1}{16} \times 45° TYP$

GEAR DATA:
ANSI B92.1-1970 FLAT ROOT~SIDE FIT

	EXTERNAL	INTERNAL
NO TEETH	34	30
DIA PITCH	4/8	4/8
PRES. ANGLE	30°	30°
PIN DIA	.480	.480
BET/OVER PINS	9.224/9.220	6.876/6.881

8D
3

6 RIBS EQUALLY SPACED
$\frac{1}{2}$" THK

$8\frac{1}{8}$
$8\frac{3}{4}$
$7\frac{1}{4}$
4
6
$\frac{1}{2}$
5
2
1
$\frac{1}{8}R$
$7\frac{1}{4}D$
$8\frac{3}{4}D$
$1\frac{3}{4}$
$1\frac{3}{4}$

MK	QUAN	DESCRIPTION	PART NO.	DWG. NO.
1	1	CONNECTING GEAR		
2	1	O-RING $\frac{1}{8} \times 1\frac{1}{4}$		
3	1	OIL IMPELLER		
4	1	GASKET .015 THK		
5	1	COLLAR $2\frac{1}{2}D \times 5/32$		
6	1	SCREW, HEX $\frac{1}{4}$-10 $\times \frac{7}{8}$		
✕		IMPELLER, ASSY.		

0 $\frac{1}{2}$ 1 2 3 4
SCALE

Problem A-3. Impeller assembly.

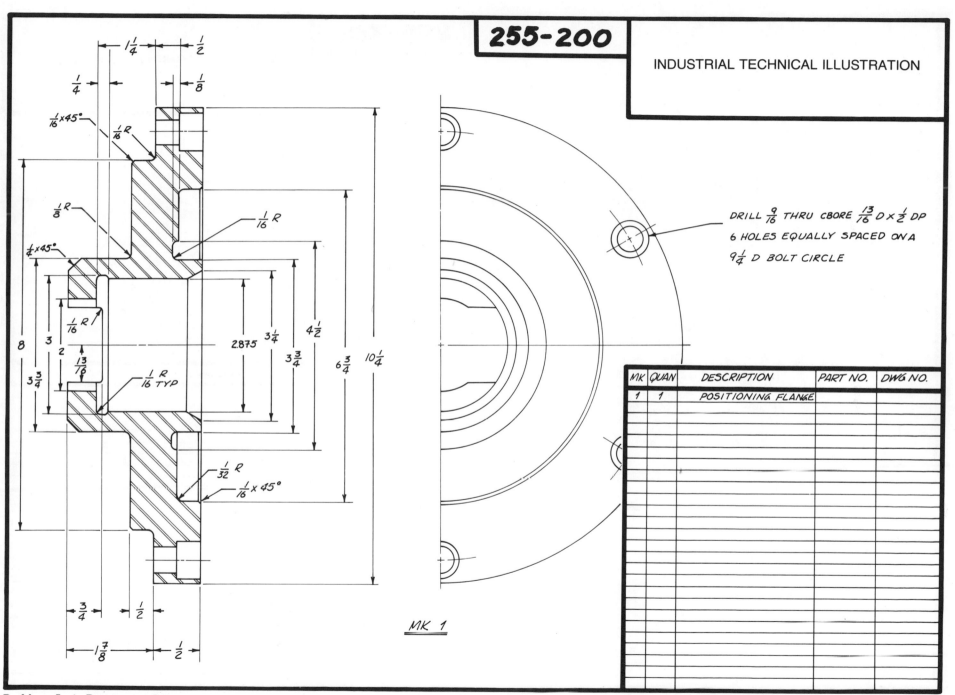

255-200

INDUSTRIAL TECHNICAL ILLUSTRATION

DRILL $\frac{9}{16}$ THRU CBORE $\frac{13}{16}$ D × $\frac{1}{2}$ DP

6 HOLES EQUALLY SPACED ON A

$9\frac{1}{4}$ D BOLT CIRCLE

MK 1

MK	QUAN	DESCRIPTION	PART NO.	DWG NO.
1	1	POSITIONING FLANGE		

Problem A-4. Positioning flange.

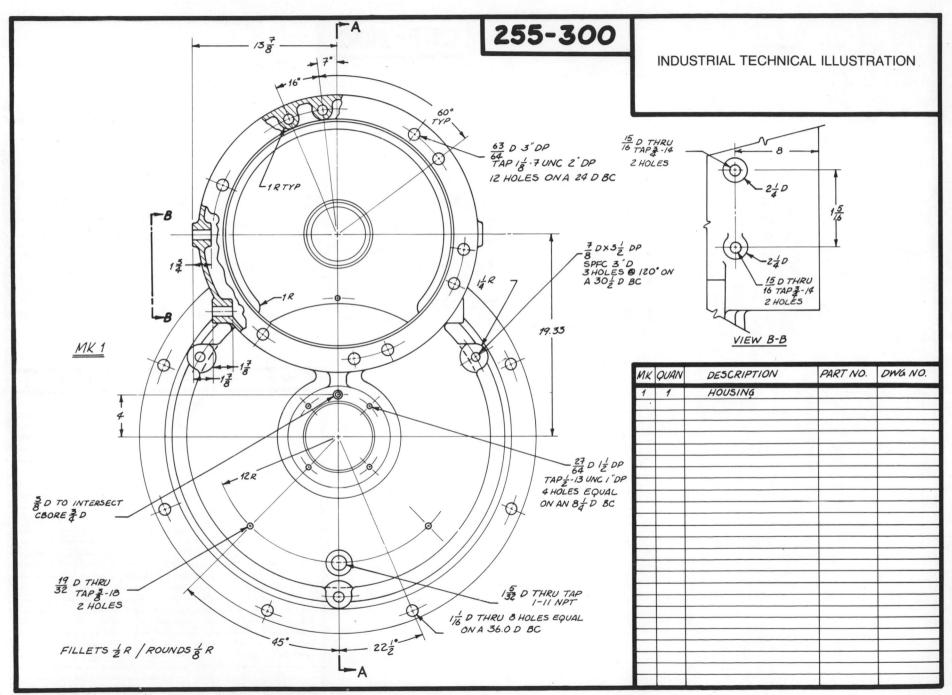

255-300

INDUSTRIAL TECHNICAL ILLUSTRATION

$13\frac{7}{8}$

$7°$

$16°$

$60°$ TYP

$\frac{63}{64}$ D 3" DP
TAP $1\frac{1}{8}$-7 UNC 2" DP
12 HOLES ON A 24 D BC

1 R TYP

B

$1\frac{3}{4}$

$\frac{1}{4}$ R

1 R

$\frac{7}{8}$ D × $3\frac{1}{2}$ DP
SPFC 3" D
3 HOLES @ 120° ON
A $30\frac{1}{2}$ D BC

19.33

$\frac{15}{16}$ D THRU
TAP $\frac{3}{4}$-14
2 HOLES

8

$2\frac{1}{4}$ D

$1\frac{5}{16}$

$2\frac{1}{4}$ D

$\frac{15}{16}$ D THRU
TAP $\frac{3}{4}$-14
2 HOLES

VIEW B-B

MK 1

$1\frac{7}{8}$

$1\frac{7}{8}$

4

$\frac{27}{64}$ D $1\frac{1}{2}$ DP
TAP $\frac{1}{2}$-13 UNC 1" DP
4 HOLES EQUAL
ON AN $8\frac{1}{4}$ D BC

12 R

$\frac{3}{8}$ D TO INTERSECT
CBORE $\frac{3}{4}$ D

$\frac{19}{32}$ D THRU
TAP $\frac{3}{8}$-18
2 HOLES

$1\frac{5}{32}$ D THRU TAP
1-11 NPT

$1\frac{1}{16}$ D THRU 8 HOLES EQUAL
ON A 36.0 D BC

45°

$22\frac{1}{2}°$

FILLETS $\frac{1}{2}$ R / ROUNDS $\frac{1}{8}$ R

MK	QUAN	DESCRIPTION	PART NO.	DWG NO.
1	1	HOUSING		

Problem A–5. Housing.

255-300

INDUSTRIAL TECHNICAL ILLUSTRATION

VIEW A-A

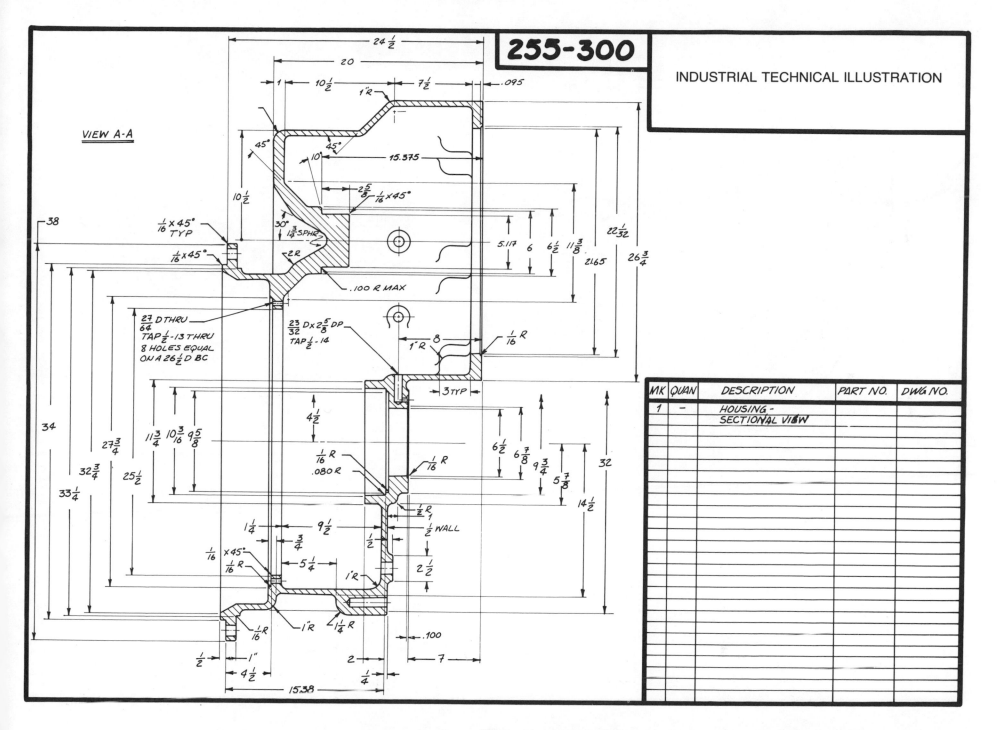

MK	QUAN	DESCRIPTION	PART NO.	DWG NO.
1	–	HOUSING - SECTIONAL VIEW		

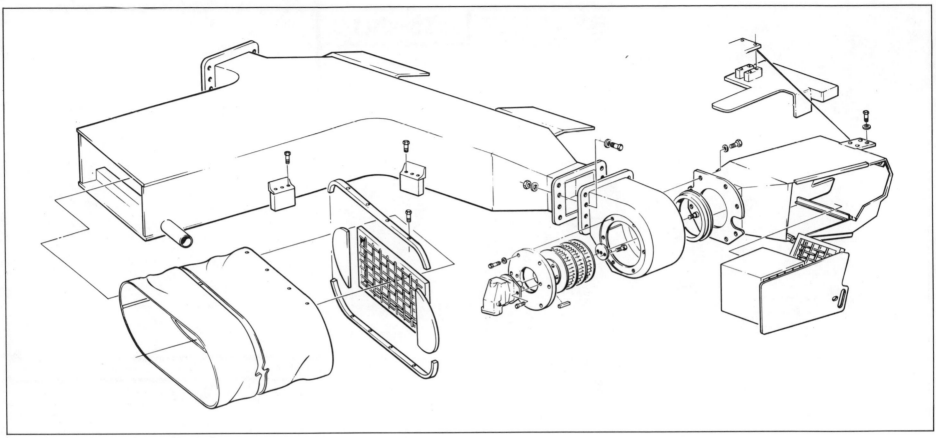

Problem A–6. Air cleaner assembly.

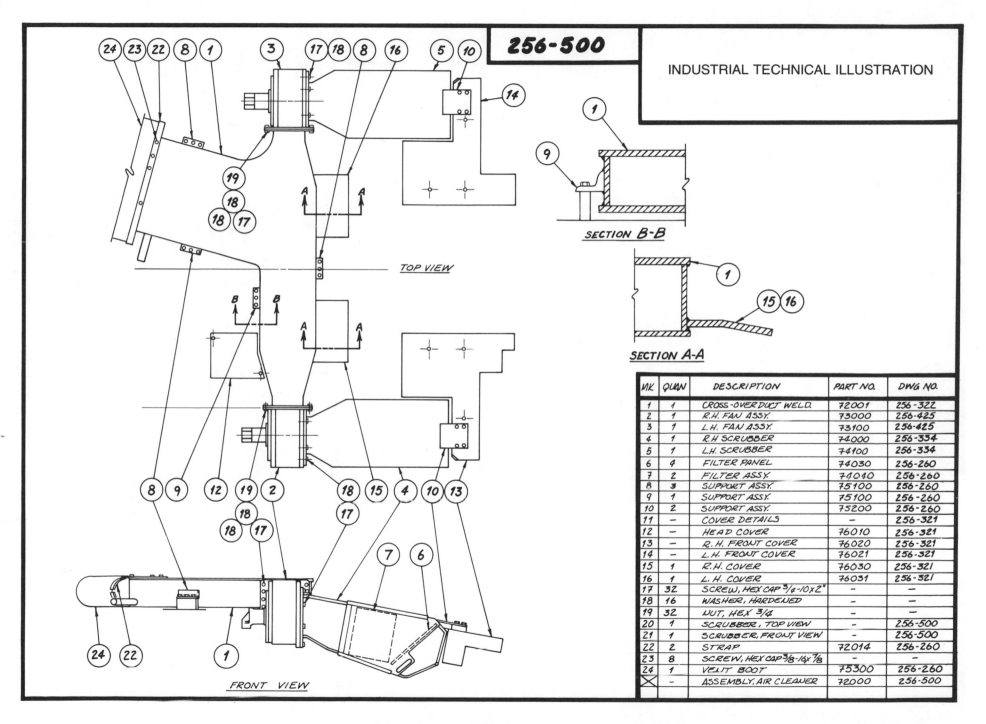

256-500

INDUSTRIAL TECHNICAL ILLUSTRATION

SECTION B-B

SECTION A-A

TOP VIEW

FRONT VIEW

MK.	QUAN	DESCRIPTION	PART NO.	DWG NO.
1	1	CROSS-OVER DUCT WELD.	72001	256-322
2	1	R.H. FAN ASSY.	73000	256-425
3	1	L.H. FAN ASSY.	73100	256-425
4	1	R.H SCRUBBER	74000	256-334
5	1	L.H. SCRUBBER	74100	256-334
6	4	FILTER PANEL	74030	256-260
7	2	FILTER ASSY.	74040	256-260
8	3	SUPPORT ASSY.	75100	256-260
9	1	SUPPORT ASSY.	75100	256-260
10	2	SUPPORT ASSY.	75200	256-260
11	–	COVER DETAILS	–	256-321
12	–	HEAD COVER	76010	256-321
13	–	R.H. FRONT COVER	76020	256-321
14	–	L.H. FRONT COVER	76021	256-321
15	1	R.H. COVER	76030	256-321
16	1	L.H. COVER	76031	256-321
17	32	SCREW, HEX CAP 3/4-10x2"	–	–
18	16	WASHER, HARDENED	–	–
19	32	NUT, HEX 3/4	–	–
20	1	SCRUBBER, TOP VIEW	–	256-500
21	1	SCRUBBER, FRONT VIEW	–	256-500
22	2	STRAP	72014	256-260
23	8	SCREW, HEX CAP 3/8-16x 7/8	–	–
24	1	VENT BOOT	75300	256-260
⊠	–	ASSEMBLY, AIR CLEANER	72000	256-500

153

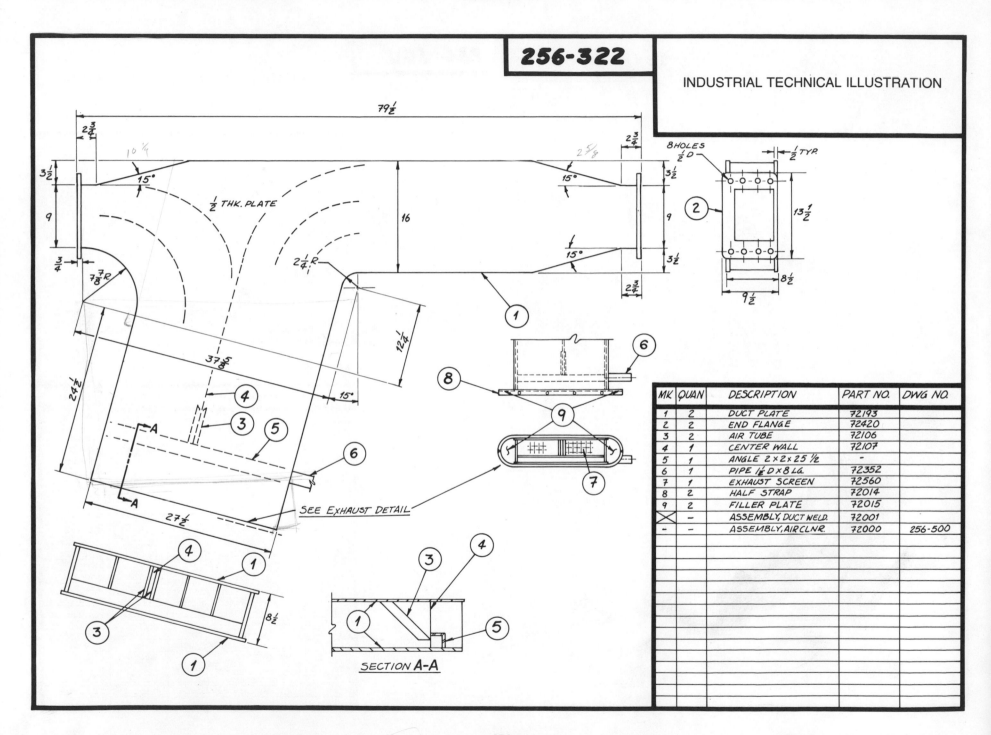

256-322

INDUSTRIAL TECHNICAL ILLUSTRATION

SEE EXHAUST DETAIL

SECTION A-A

MK	QUAN	DESCRIPTION	PART NO.	DWG NO.
1	2	DUCT PLATE	72193	
2	2	END FLANGE	72420	
3	2	AIR TUBE	72106	
4	1	CENTER WALL	72107	
5	1	ANGLE 2 x 2 x 25 1/2	-	
6	1	PIPE 1/2 D x 8 LG.	72352	
7	1	EXHAUST SCREEN	72560	
8	2	HALF STRAP	72014	
9	2	FILLER PLATE	72015	
X	-	ASSEMBLY, DUCT WELD.	72001	
-	-	ASSEMBLY, AIR CLNR.	72000	256-500

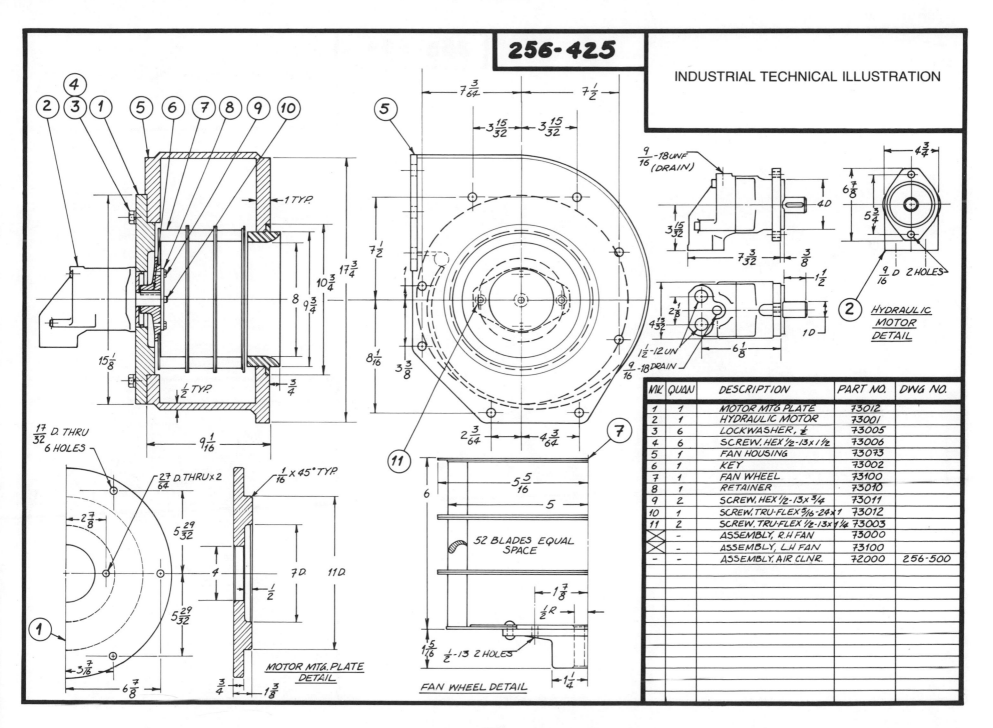

256-425

INDUSTRIAL TECHNICAL ILLUSTRATION

HYDRAULIC MOTOR DETAIL

MOTOR MTG. PLATE DETAIL

FAN WHEEL DETAIL

52 BLADES EQUAL SPACE

MK.	QUAN.	DESCRIPTION	PART NO.	DWG NO.
1	1	MOTOR MTG PLATE	73012	
2	1	HYDRAULIC MOTOR	73001	
3	6	LOCKWASHER, 1/2	73005	
4	6	SCREW, HEX 1/2-13 x 1 1/2	73006	
5	1	FAN HOUSING	73073	
6	1	KEY	73002	
7	1	FAN WHEEL	73100	
8	1	RETAINER	73010	
9	2	SCREW, HEX 1/2-13 x 3/4	73011	
10	1	SCREW, TRU-FLEX 5/16-24 x 1	73012	
11	2	SCREW, TRU-FLEX 1/2-13 x 1 1/4	73003	
⊠	–	ASSEMBLY, R.H FAN	73000	
⊠	–	ASSEMBLY, LH FAN	73100	
–	–	ASSEMBLY, AIR CLNR.	72000	256-500

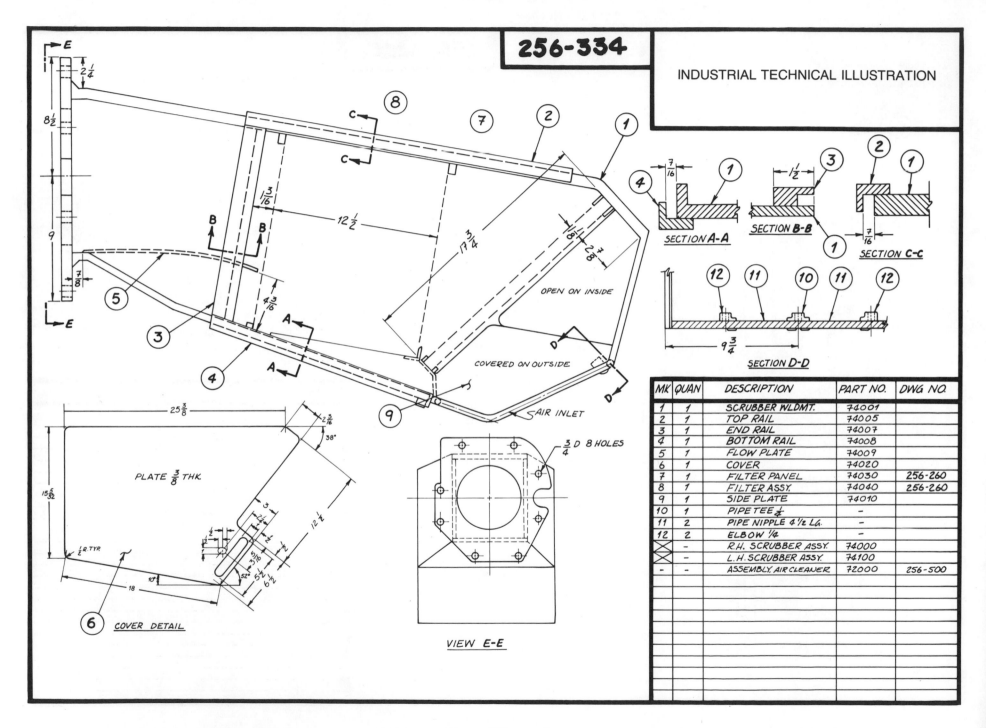

256-334

INDUSTRIAL TECHNICAL ILLUSTRATION

SECTION A-A

SECTION B-B

SECTION C-C

SECTION D-D

OPEN ON INSIDE

COVERED ON OUTSIDE

AIR INLET

PLATE 3/8 THK.

COVER DETAIL

3/4 D 8 HOLES

VIEW E-E

MK	QUAN	DESCRIPTION	PART NO.	DWG NO.
1	1	SCRUBBER WLDMT.	74001	
2	1	TOP RAIL	74005	
3	1	END RAIL	74007	
4	1	BOTTOM RAIL	74008	
5	1	FLOW PLATE	74009	
6	1	COVER	74020	
7	1	FILTER PANEL	74030	256-260
8	1	FILTER ASSY.	74040	256-260
9	1	SIDE PLATE	74010	
10	1	PIPE TEE 1/4	–	
11	2	PIPE NIPPLE 4 1/2 LG.	–	
12	2	ELBOW 1/4	–	
✕	–	R.H. SCRUBBER ASSY.	74000	
✕	–	L.H. SCRUBBER ASSY.	74100	
–	–	ASSEMBLY, AIR CLEANER	72000	256-500

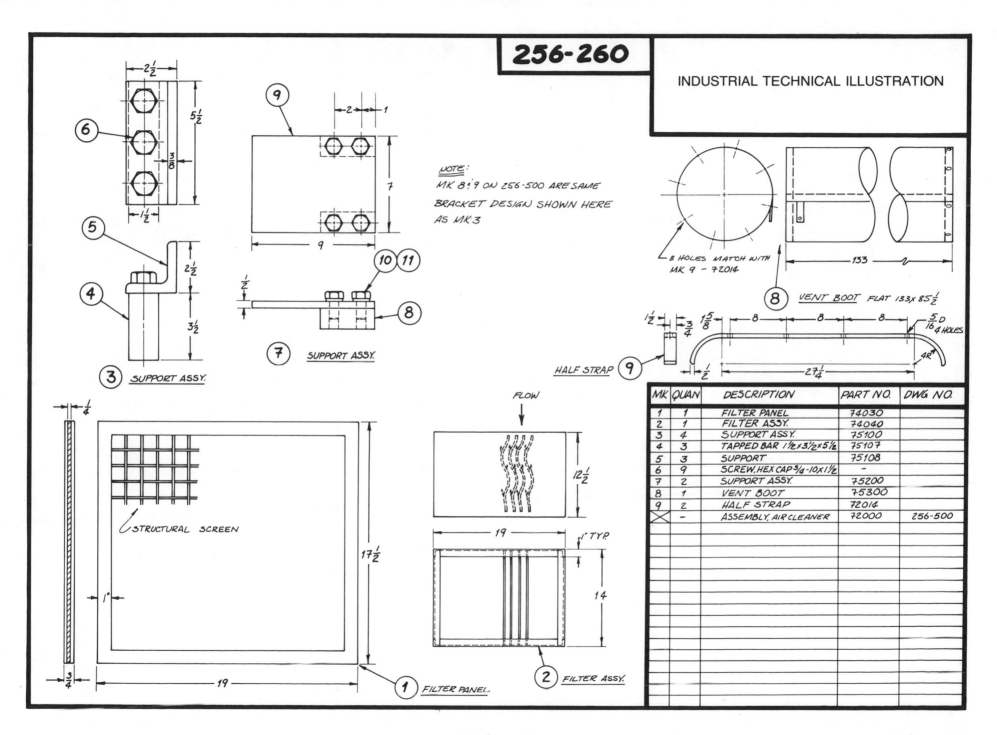

256-260

INDUSTRIAL TECHNICAL ILLUSTRATION

NOTE:
MK 8 & 9 ON 256-500 ARE SAME
BRACKET DESIGN SHOWN HERE
AS MK·3

8 HOLES MATCH WITH
MK 9 - 72014

VENT BOOT FLAT 133 x 85½

③ SUPPORT ASSY.

⑦ SUPPORT ASSY.

HALF STRAP

STRUCTURAL SCREEN

FLOW

① FILTER PANEL.

② FILTER ASSY.

MK	QUAN	DESCRIPTION	PART NO.	DWG NO.
1	1	FILTER PANEL	74030	
2	1	FILTER ASSY.	74040	
3	4	SUPPORT ASSY.	75100	
4	3	TAPPED BAR 1½x3½x5½	75107	
5	3	SUPPORT	75108	
6	9	SCREW, HEX CAP ¾-10x1½	—	
7	2	SUPPORT ASSY.	75200	
8	1	VENT BOOT	75300	
9	2	HALF STRAP	72014	
✕	—	ASSEMBLY, AIR CLEANER	72000	256-500

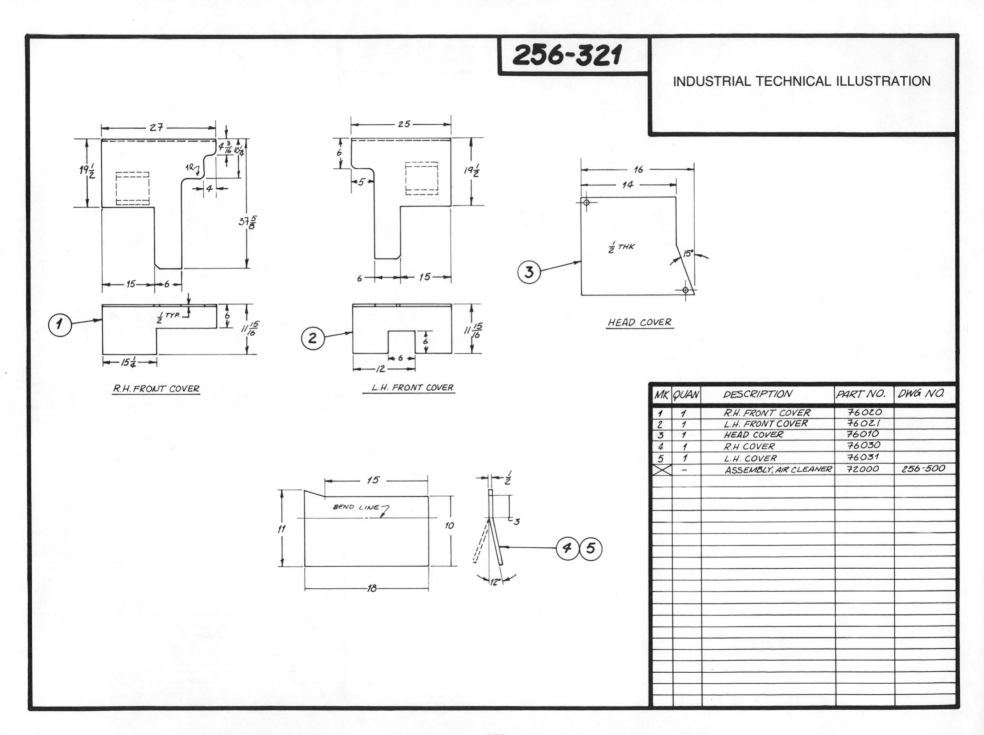

256-321

INDUSTRIAL TECHNICAL ILLUSTRATION

R.H. FRONT COVER

L.H. FRONT COVER

HEAD COVER

BEND LINE

MK	QUAN	DESCRIPTION	PART NO.	DWG NO.
1	1	R.H. FRONT COVER	76020	
2	1	L.H. FRONT COVER	76021	
3	1	HEAD COVER	76010	
4	1	R.H. COVER	76030	
5	1	L.H. COVER	76031	
X	–	ASSEMBLY, AIR CLEANER	72000	256-500

Problem A–7. Control pedant.

Problem A–8. Power transformer.

Problem A–9. Cable connection.

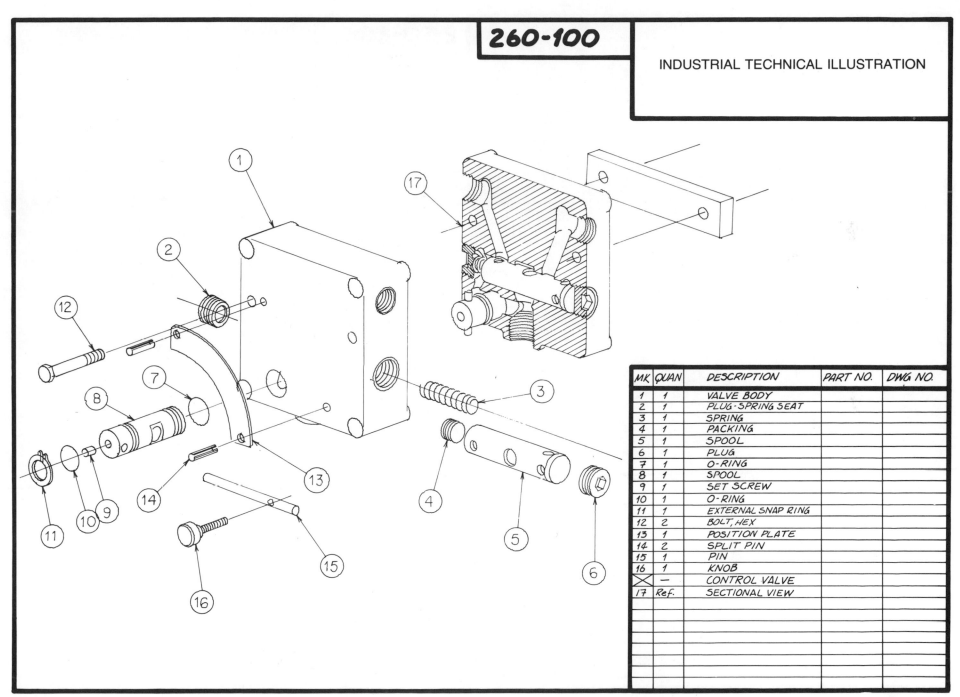

260-100

INDUSTRIAL TECHNICAL ILLUSTRATION

MK	QUAN	DESCRIPTION	PART NO.	DWG NO.
1	1	VALVE BODY		
2	1	PLUG-SPRING SEAT		
3	1	SPRING		
4	1	PACKING		
5	1	SPOOL		
6	1	PLUG		
7	1	O-RING		
8	1	SPOOL		
9	1	SET SCREW		
10	1	O-RING		
11	1	EXTERNAL SNAP RING		
12	2	BOLT, HEX		
13	1	POSITION PLATE		
14	2	SPLIT PIN		
15	1	PIN		
16	1	KNOB		
✕	—	CONTROL VALVE		
17	Ref.	SECTIONAL VIEW		

Problem A–10. Control valve.

260-200

INDUSTRIAL TECHNICAL ILLUSTRATION

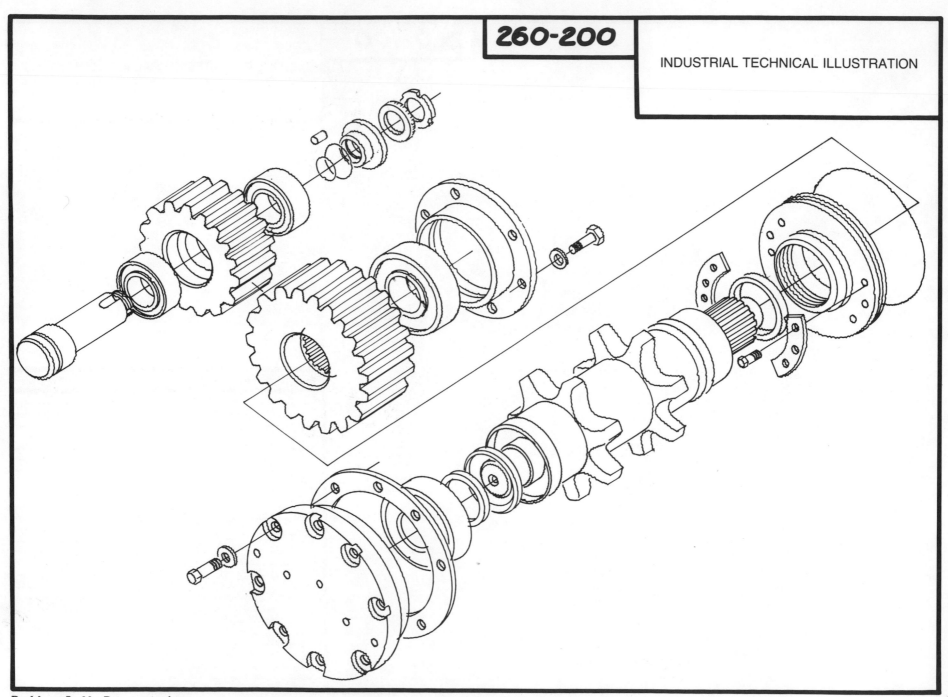

Problem A–11. Drive sprocket.

APPENDIX B

Fasteners and Equipment

This appendix presents many of the standard parts you will encounter as an illustrator. They have been drawn in line technique for your reference. You should become familiar with both their names and their shapes. The more common parts have been shown in their construction and in their standard sizes. You should memorize these parts in order to spend as little time as possible in construction. Even when you are using pressure-sensitive transfers, your knowledge of construction will aid in choosing the correct transfer. When working the problems in Appendix A, keep this section as a reference.

B-1. Ball bearing.

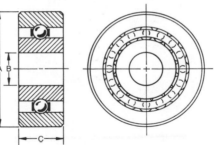

CONSTRUCTION   FINISHED

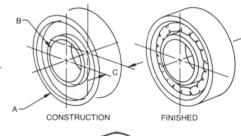

Nominal A	Nominal B	Nominal C
2.00	1.00	.60
2.50	1.20	.63
3.00	1.50	.70
3.50	2.00	.78
4.00	2.75	.82
4.75	2.20	1.15
5.00	3.36	1.25
5.50	2.16	1.30
6.00	2.40	1.37

B-2. Common spring.

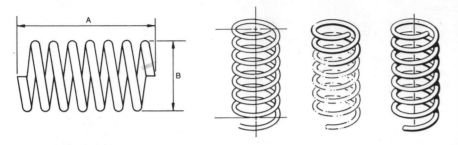

B-3. Cotter pin..

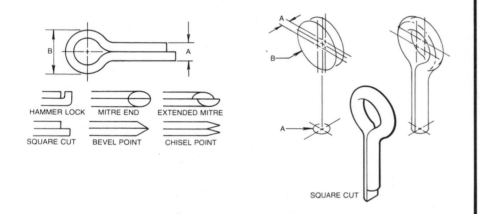

B-5. Hexagonal head cap screw.

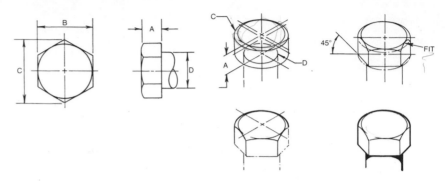

Nominal Size D	Across Flats B	Across Corners C	Head Height A
1/4	7/16	.505	5/32
5/16	1/2	.577	13/64
3/8	9/16	.650	15/64
7/16	5/8	.722	9/32
1/2	3/4	.866	5/16
9/16	13/16	.938	23/64
5/8	15/16	1.083	25/64
3/4	1 1/8	1.299	15/32
7/8	1 5/16	1.516	35/64
1	1 1/2	1.732	39/64
1 1/8	1 11/16	1.949	11/16
1 1/4	2 1/16	2.382	25/32

B-4. Flat head cap screw.

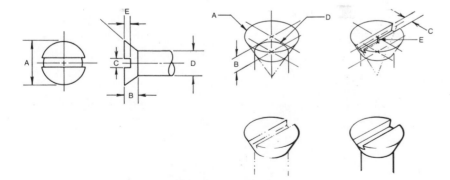

Nominal Size D	Head Diameter A	Head Height B	Slot Width C	Slot Depth E
1/4	.500	.140	.075	.069
5/16	.625	.176	.084	.086
3/8	.750	.210	.094	.103
7/16	.813	.210	.094	.103
1/2	.875	.210	.106	.103
9/16	1.000	.245	.118	.120
5/8	1.125	.281	.133	.137
3/4	1.375	.352	.149	.171
7/8	1.625	.423	.167	.206
1	1.875	.494	.188	.240

B–6. Hexagonal nut.

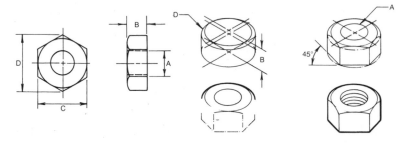

Nominal Size A	Head Thickness B	Across Flats C	Across Corners D
1/4	5/32	7/16	.505
5/16	3/16	1/2	.577
3/8	7/32	9/16	.650
7/16	1/4	11/16	.794
1/2	5/16	3/4	.866
9/16	5/16	7/8	1.010
5/8	3/8	15/16	1.083
3/4	27/64	1 1/8	1.299
7/8	31/64	1 5/16	1.516
1	35/64	1 1/2	1.732
1 1/4	27/32	2 1/4	2.598

B–7. Holes.

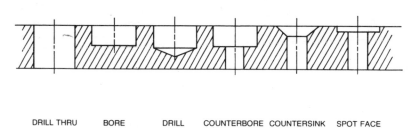

DRILL THRU BORE DRILL COUNTERBORE COUNTERSINK SPOT FACE

B–8. Hose clamp.

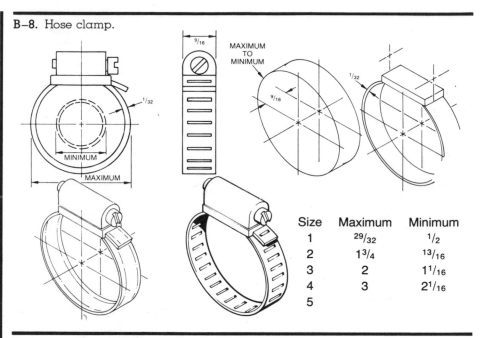

Size	Maximum	Minimum
1	29/32	1/2
2	1 3/4	13/16
3	2	1 1/16
4	3	2 1/16
5		

B–9. Keys and key seats.

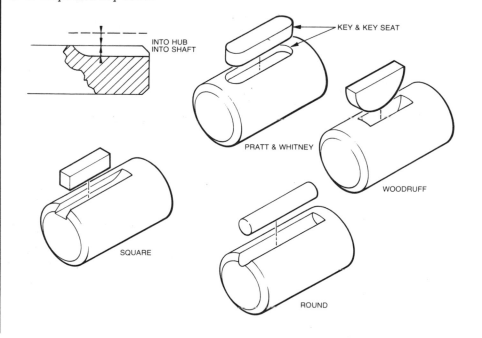

INTO HUB
INTO SHAFT

KEY & KEY SEAT

PRATT & WHITNEY

WOODRUFF

SQUARE

ROUND

B–10. Knurls.

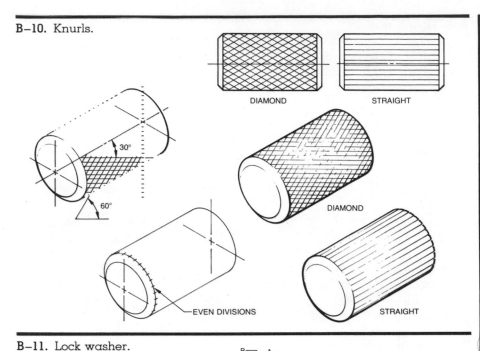

DIAMOND

STRAIGHT

30°

60°

DIAMOND

EVEN DIVISIONS

STRAIGHT

B–12. Lube fitting.

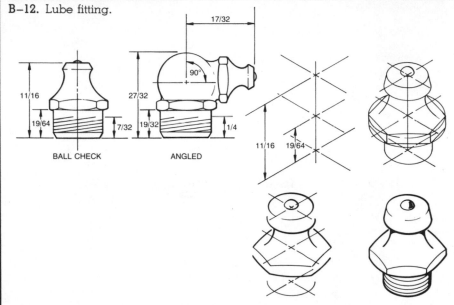

17/32

90°

11/16

19/64

7/32

27/32

19/32

1/4

11/16

19/64

BALL CHECK

ANGLED

B–11. Lock washer.

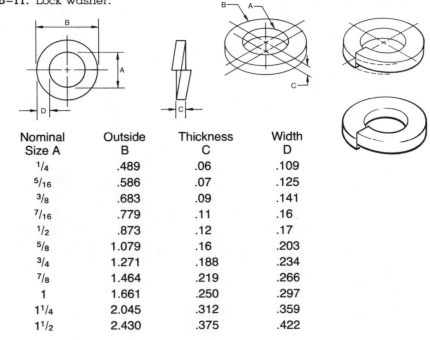

B

A

C

D

Nominal Size A	Outside B	Thickness C	Width D
1/4	.489	.06	.109
5/16	.586	.07	.125
3/8	.683	.09	.141
7/16	.779	.11	.16
1/2	.873	.12	.17
5/8	1.079	.16	.203
3/4	1.271	.188	.234
7/8	1.464	.219	.266
1	1.661	.250	.297
1 1/4	2.045	.312	.359
1 1/2	2.430	.375	.422

B–13. O-rings.

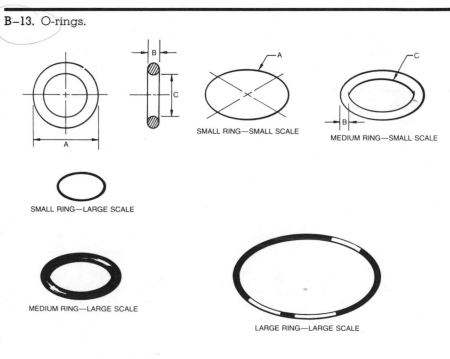

A

B

C

A

B

C

SMALL RING—SMALL SCALE

MEDIUM RING—SMALL SCALE

SMALL RING—LARGE SCALE

MEDIUM RING—LARGE SCALE

LARGE RING—LARGE SCALE

B-14. Rings, springs, and bearings.

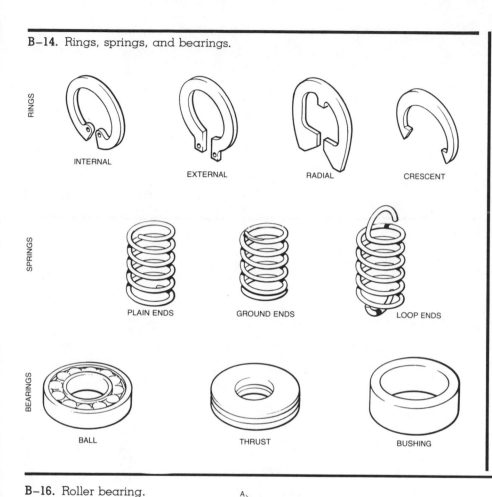

RINGS

INTERNAL   EXTERNAL   RADIAL   CRESCENT

SPRINGS

PLAIN ENDS   GROUND ENDS   LOOP ENDS

BEARINGS

BALL   THRUST   BUSHING

B-15. Roll pin.

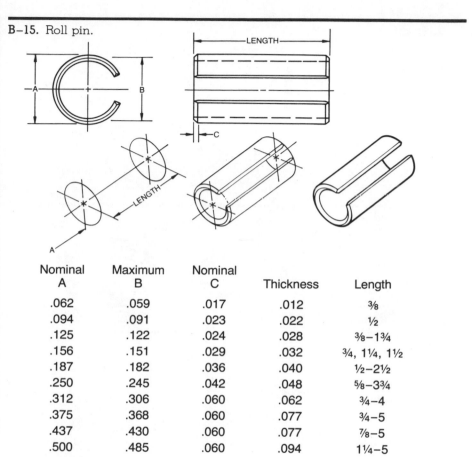

Nominal A	Maximum B	Nominal C	Thickness	Length
.062	.059	.017	.012	⅜
.094	.091	.023	.022	½
.125	.122	.024	.028	⅜–1¾
.156	.151	.029	.032	¾, 1¼, 1½
.187	.182	.036	.040	½–2½
.250	.245	.042	.048	⅝–3¾
.312	.306	.060	.062	¾–4
.375	.368	.060	.077	¾–5
.437	.430	.060	.077	⅞–5
.500	.485	.060	.094	1¼–5

B-16. Roller bearing.

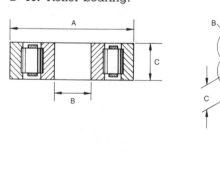

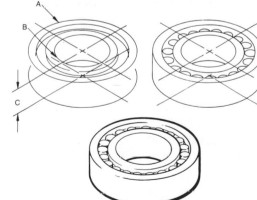

Nominal A	Nominal B	Nominal C
1.00	2.44	.669
1.37	4.62	.984
1.77	3.34	1.187
1.96	4.33	1.063
2.55	4.72	1.50
2.75	7.08	1.65
2.95	5.11	.984

B–17. Round head cap screw.

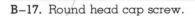
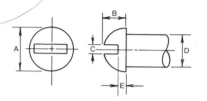
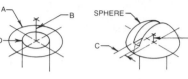
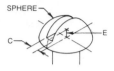

Nominal Size D	Head Diameter A	Head Height B	Slot Width C	Slot Depth E
1/4	.437	.191	.075	.117
5/16	.562	.246	.084	.151
3/8	.625	.273	.094	.168
7/16	.750	.328	.094	.202
1/2	.812	.355	.106	.219
9/16	.937	.410	.118	.253
5/8	1.000	.438	.133	.270
3/4	1.250	.547	.149	.337

B–18. Seals and gears.

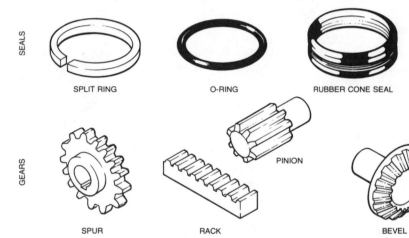

SEALS

SPLIT RING   O-RING   RUBBER CONE SEAL

GEARS

SPUR   RACK   PINION   BEVEL

B–20. Set screws.

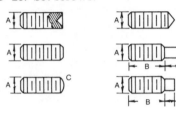
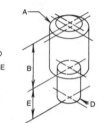
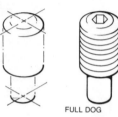

FULL DOG

Nominal Size A	Length B	Oval Point Radius C	Dog Point Diameter D	Full Dog E	Half Dog e
1/4	5/16	.188	.156	.125	.063
3/8	7/16	.281	.250	.188	.094
1/2	9/16	.375	.344	.250	.125
5/8	11/16	.469	.469	.313	.156
3/4	7/8	.656	.658	.458	.219
1	1 1/8	.750	.750	.500	.250

B–19. Self-tapping screws.

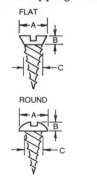
FLAT   PAN
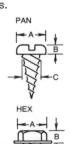
ROUND   HEX
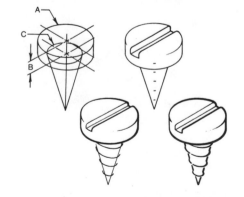

Nominal Size	Flat			Round			Pan			Hex		
	A	B	C	A	B	C	A	B	C	A	B	C
0	.119	.035	.060	.113	.053	.060	.119	.035	.060	.119	.035	.060
2	.172	.051	.086	.162	.069	.086	.172	.051	.086	.172	.051	.086
4	.225	.067	.112	.211	.086	.112	.225	.067	.112	.225	.067	.112
6	.279	.083	.138	.260	.103	.138	.279	.083	.138	.279	.083	.138
8	.332	.100	.164	.309	.120	.164	.332	.100	.164	.332	.100	.164
10	.385	.116	.190	.359	.137	.190	.385	.116	.190	.385	.116	.190

FASTENERS AND EQUIPMENT

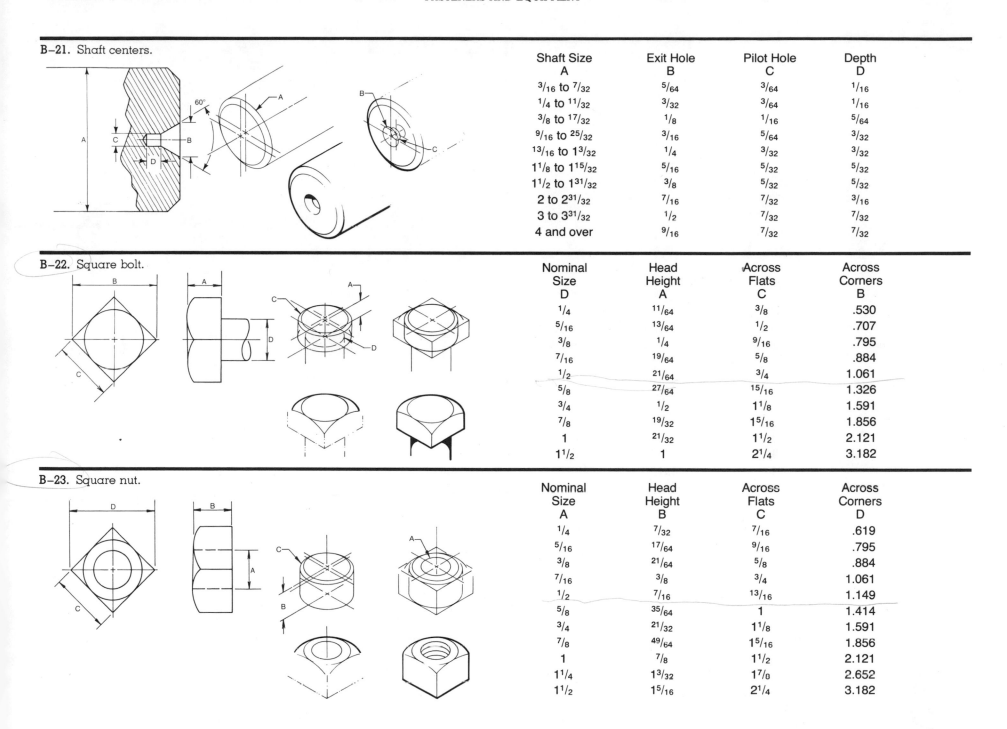

B-21. Shaft centers.

Shaft Size A	Exit Hole B	Pilot Hole C	Depth D
$3/16$ to $7/32$	$5/64$	$3/64$	$1/16$
$1/4$ to $11/32$	$3/32$	$3/64$	$1/16$
$3/8$ to $17/32$	$1/8$	$1/16$	$5/64$
$9/16$ to $25/32$	$3/16$	$5/64$	$3/32$
$13/16$ to $13/32$	$1/4$	$3/32$	$3/32$
$11/8$ to $115/32$	$5/16$	$5/32$	$5/32$
$11/2$ to $131/32$	$3/8$	$5/32$	$5/32$
2 to $231/32$	$7/16$	$7/32$	$3/16$
3 to $331/32$	$1/2$	$7/32$	$7/32$
4 and over	$9/16$	$7/32$	$7/32$

B-22. Square bolt.

Nominal Size D	Head Height A	Across Flats C	Across Corners B
$1/4$	$11/64$	$3/8$	.530
$5/16$	$13/64$	$1/2$	.707
$3/8$	$1/4$	$9/16$	.795
$7/16$	$19/64$	$5/8$	.884
$1/2$	$21/64$	$3/4$	1.061
$5/8$	$27/64$	$15/16$	1.326
$3/4$	$1/2$	$11/8$	1.591
$7/8$	$19/32$	$15/16$	1.856
1	$21/32$	$11/2$	2.121
$11/2$	1	$21/4$	3.182

B-23. Square nut.

Nominal Size A	Head Height B	Across Flats C	Across Corners D
$1/4$	$7/32$	$7/16$	.619
$5/16$	$17/64$	$9/16$	.795
$3/8$	$21/64$	$5/8$	.884
$7/16$	$3/8$	$3/4$	1.061
$1/2$	$7/16$	$13/16$	1.149
$5/8$	$35/64$	1	1.414
$3/4$	$21/32$	$11/8$	1.591
$7/8$	$49/64$	$15/16$	1.856
1	$7/8$	$11/2$	2.121
$11/4$	$13/32$	$17/0$	2.652
$11/2$	$15/16$	$21/4$	3.182

B–24. Snap ring.

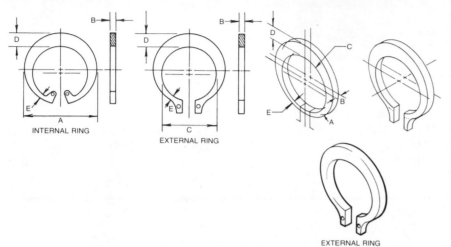

INTERNAL RING

EXTERNAL RING

EXTERNAL RING

B–25. Socket head cap screw.

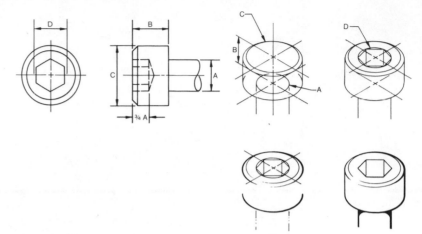

Nominal A	Internal Ring Thickness B	Offset D	Neck E
1/4	.015	.025	.015
3/8	.025	.040	.028
7/16	.025	.049	.029
1/2	.035	.053	.035
5/8	.035	.060	.035
3/4	.035	.070	.040
7/8	.042	.084	.045
1	.042	.155	.052

Nominal C	External Ring Thickness B	Offset D	Neck E
1/4	.025	.035	.025
3/8	.025	.050	.030
7/16	.025	.055	.033
1/2	.035	.065	.040
5/8	.035	.080	.045
3/4	.042	.092	.051
7/8	.042	.104	.057
1	.042	.116	.065

Nominal A	Thickness B	Head C	Socket D
1/4	1/4	3/8	3/16
5/16	5/16	7/16	7/32
3/8	3/8	9/16	5/16
7/16	7/16	5/8	5/16
1/2	1/2	3/4	3/8
9/16	9/16	13/16	3/8
5/8	5/8	7/8	1/2
3/4	3/4	1	9/16
7/8	7/8	1 1/8	9/16
1	1	1 5/16	5/8
1 1/4	1 1/4	1 3/4	3/4
1 1/2	1 1/2	2	1

B–26. Special bolts.

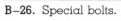
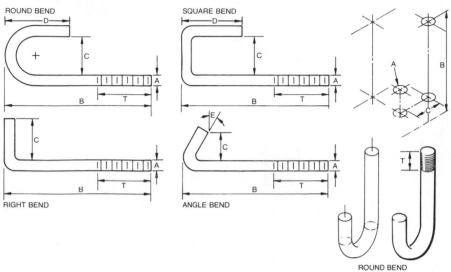

ROUND BEND

SQUARE BEND

RIGHT BEND

ANGLE BEND

ROUND BEND

A	Round			Square			Right			Angle			
	B	C	D	B	C	D	B	C	D	B	C	D	E
3/8-16	1 9/32	9/16	17/32										
1/2-13	1 3/4	1/2	5/8				1 1/2	2		7		15°	
5/8-11							1 1/4	2					
#10-24				2 1/16	5/16	1 3/16							

B–27. Splines and hubs.

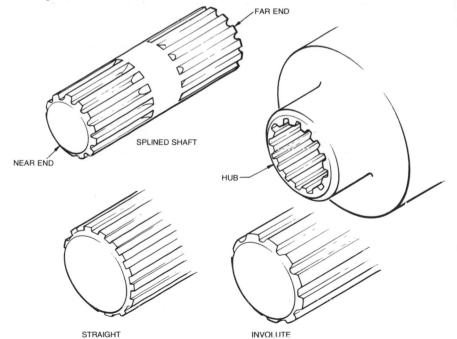

FAR END

SPLINED SHAFT

NEAR END

HUB

STRAIGHT

INVOLUTE

B–28. Taper pin.

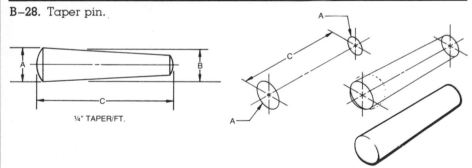

1/4" TAPER/FT.

Size	Large End A	Drill Size B	Maximum Length C
00	0.147	31	1
0	0.156	28	1
1	0.172	25	1 1/4
2	0.193	19	1 1/2
3	0.219	12	1 3/4
4	0.250	3	2
6	0.341	9/32	3 1/4
8	0.492	13/32	4 1/2
10	0.706	19/32	6
12	1.013	55/64	8 3/4

B–29. Threads, heads, and drives.

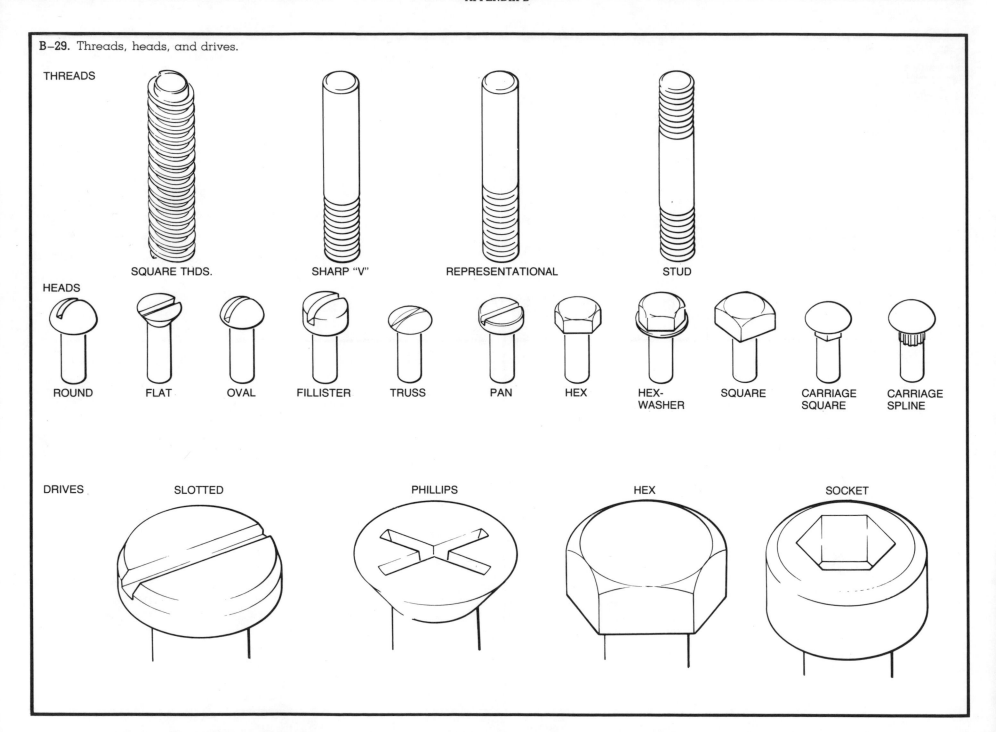

THREADS

SQUARE THDS.   SHARP "V"   REPRESENTATIONAL   STUD

HEADS

ROUND   FLAT   OVAL   FILLISTER   TRUSS   PAN   HEX   HEX-WASHER   SQUARE   CARRIAGE SQUARE   CARRIAGE SPLINE

DRIVES   SLOTTED   PHILLIPS   HEX   SOCKET

B-30. Thrust bearing.

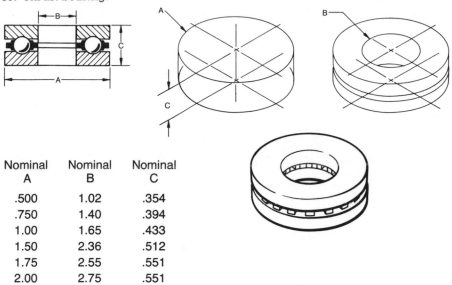

Nominal A	Nominal B	Nominal C
.500	1.02	.354
.750	1.40	.394
1.00	1.65	.433
1.50	2.36	.512
1.75	2.55	.551
2.00	2.75	.551
2.50	3.54	.709

B-31. Thumb screw.

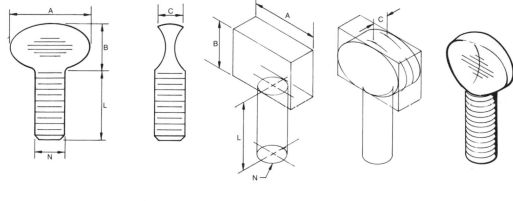

Nominal N	Head W. A	Head H. B	Thickness C	Length L
$1/4$	.813	.500	.115	$1/2$
$5/16$	.938	.625	.156	$5/8$
$3/8$	1.063	.688	.200	$3/4$
$7/16$	1.375	.938	.258	$1 1/4$
$1/2$	1.500	1.013	.312	$1 1/2$

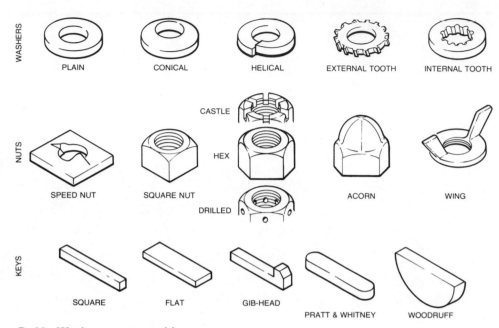

WASHERS

PLAIN CONICAL HELICAL EXTERNAL TOOTH INTERNAL TOOTH

NUTS

SPEED NUT SQUARE NUT CASTLE HEX DRILLED ACORN WING

KEYS

SQUARE FLAT GIB-HEAD PRATT & WHITNEY WOODRUFF

B–32. Washers, nuts, and keys.

APPENDIX C
Tables

Picking an inappropriate technique to render an illustration can make an easy job twice as difficult as it might be. Line illustration is the standard for all materials but, as you can see, other techniques may be more appropriate for a specific material.

To reproduce illustration most effectively, drawing technique should be matched to the reproduction methods available.

Use Tables C-4 and C-5 to make your own scales or projection diagrams for some of the more popular axonometric views.

Table C-1. Best Rendering Technique

Material	Technique			
	Line	Block	Stipple	Airbrush
Cast metal	✓		✓	
Machined metal	✓	✓		
Formed metal	✓			
Plastic	✓			
Glass	✓			✓
Rubber	✓	✓	✓	✓
Cloth	✓		✓	✓
Skin	✓		✓	✓

Table C-2. Best Reproduction Method

Rendering Technique	Reproduction				
	Ozalid	Photo	Halftone	Microfilm	Electronic
Line Shading	✓	✓			✓
Block Shading	✓	✓		✓	
Stippling	✓	✓			
Airbrush		✓	✓		

Table C-3. Decimal Conversion

Fraction	Decimal	Fraction	Decimal	Fraction	Decimal
1/64	.0156	23/64	.3594	45/64	.7031
1/32	.0312	3/8	.3750	23/32	.7188
3/64	.0469	27/64	.3906	47/64	.7344
1/16	.0625	13/32	.4062	3/4	.7500
5/64	.0781	27/64	.4219	49/64	.7656
3/32	.0938	7/16	.4375	25/32	.7812
7/64	.1094	29/64	.4531	51/64	.7969
1/8	.1250	15/32	.4688	13/16	.8125
9/64	.1406	31/64	.4844	53/64	.8281
5/32	.1562	1/2	.5000	27/32	.8438
11/64	.1719	33/64	.5156	55/64	.8594
3/16	.1875	17/32	.5312	7/8	.8750
13/64	.2031	35/64	.5469	57/64	.8906
7/32	.2188	9/16	.5625	29/32	.9062
15/64	.2344	37/64	.5781	59/64	.9219
1/4	.2500	19/32	.5938	15/16	.9375
17/64	.2656	39/64	.6094	61/64	.9531
9/32	.2812	5/8	.6250	31/32	.9688
19/64	.2969	41/64	.6406	63/64	.9844
5/16	.3125	21/32	.6562	1	1.0000
21/64	.3281	43/64	.6719		
11/32	.3438	11/16	.6875		

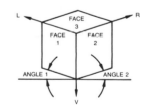

Table C-4. Dimetric Scales and Ellipses

View Name	Scaling Factor			Angles		Ellipses in Faces		
	V	R	L	1	2	1	2	3
20–20–60	.500	1.00	1.00	40°	40°	20°	20°	60°
25–25–50	.625	1.00	1.00	39°	39°	25°	25°	50°
30–30–50	.750	1.00	1.00	37°	37°	30°	30°	50°
40–40–20	.810	.81	1.00	20°	20°	40°	40°	20°
45–45–15	1.000	.75	.75	15°	15°	45°	45°	15°

When you need to determine the correct ellipse required for a particular application, knowledge of the major and minor axis ratio and this table will give you the answer. Often you can construct a unitary (equal-sided) box that the ellipse would fit in. Either visually fit by trial and error or figure the ellipse using the axis ratio.

Table C-5. Trimetric Scales and Ellipses

View Name	Scaling Factor			Angles		Ellipses in Faces		
	V	R	L	1	2	1	2	3
55–35–15	.96	.60	.84	10°	20°	55°	35°	15°
55–25–20	.94	.56	.88	10°	30°	55°	25°	20°
45–40–20	.95	.68	.78	15°	20°	45°	40°	20°
50–30–25	.86	.65	.92	15°	30°	50°	30°	25°
45–35–25	.88	.72	.84	20°	30°	45°	35°	25°

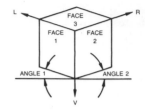

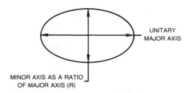

UNITARY MAJOR AXIS

MINOR AXIS AS A RATIO OF MAJOR AXIS (R)

Table C-6. Axis Ratio of Non-available Ellipses

Ellipse Desired	Minor Axis as a % of a Major Axis	Ellipse Desired	Minor Axis as a % of a Major Axis	Ellipse Desired	Minor Axis as a % of a Major Axis	Ellipse Desired	Minor Axis as a % of a Major Axis	Ellipse Desired	Minor Axis as a % of a Major Axis
1°	.017	19°	.326	37°	.602	55°	available	73°	.956
2°	.035	20°	available	38°	.616	56°	.829	74°	.961
3°	.052	21°	.358	39°	.629	57°	.838	75°	available
4°	.070	22°	.375	40°	available	58°	.848	76°	.970
5°	available	23°	.390	41°	.656	59°	.857	77°	.974
6°	.104	24°	.407	42°	.669	60°	available	78°	.978
7°	.123	25°	available	43°	.682	61°	.875	79°	.981
8°	.142	26°	.438	44°	.695	62°	.883	80°	available
9°	.156	27°	.454	45°	available	63°	.891	81°	.987
10°	available	28°	.469	46°	.719	64°	.899	82°	.990
11°	.191	29°	.485	47°	.731	65°	available	83°	.992
12°	.208	30°	available	48°	.743	66°	.913	84°	.994
13°	.225	31°	.515	49°	.755	67°	.902	85°	.996
14°	.242	32°	.529	50°	available	68°	.927	86°	.997
15°	available	33°	.545	51°	.777	69°	.934	87°	.998
16°	.276	34°	.559	52°	.788	70°	available	88°	.9993
17°	.292	35°	available	53°	.798	71°	.945	89°	.9998
18°	.309	36°	.587	54°	.809	72°	.951		

Terms and Abbreviations

Many engineering drawings contain abbreviations for common terms. The abbreviations usually require less time to produce and aid in the rapid reading of the drawing. If an abbreviation is unique to a particular industry or company and the drawing is to be used by someone outside that area, the term should be defined on the drawing or appropriately referenced to a glossary.

Term	Abbreviation	Term	Abbreviation	Term	Abbreviation	Term	Abbreviation
Adapter	ADPT.	Degree	DEG	Keyway	KWY	Quantity	QTY
Adjust	ADJ.	Diagram	DIAG	Left hand	LH	Rectangle	RECT
Amount	AMT	Dimension	DIM.	Length overall	LOA	Reference	REF
Approved	APPD	Drawing list	DL	Long	LG	Reference line	REF L
Approximate	APPROX	Each	EA	Machine	MACH	Required	REQD
Area	A	End to end	EtoE	Manual	MAN.	Right hand	RH
Assemble	ASSEM	Exhaust	EXH	Material	MATL	Round	RD
Assembly	ASSY	Exterior	EXT	Material list	ML	Schematic	SCHEM
Attach	ATT	Face to face	FtoF	Maximum	MAX	Screw	SCR
Back to back	BtoB	Figure	FIG.	Miscellaneous	MISC	Shaft	SFT
Base line	BL	Fillet	FIL	Model	MOD	Specification	SPEC
Between centers	BC	Finish	FIN	Mounted	MTD	Square	SQ
Bill of materials	BM	Flat	F	Nominal	NOM	Standard	STD
Blueprint	BP	Flat head	FH	Not to scale	NTS	Straight	STR
Cap screw	CAP SCR	Foot	(') FT	Number	NO.	Surface	SUR
Castle nut	CAS NUT	Gasket	GSKT	Octagon	OCT	Tangent	TAN.
Center	CTR	Grade	GR	Outlet	OUT.	Template	TEMP
Center line	CL	Graphic	GRAPH.	Overall	OA	Thick	THK
Center to center	CtoC	Ground	GRD	Overhead	OVHD	Thread	THD
Chamfer	CHAM	Head	HD	Perpendicular	PERP	Tolerance	TOL
Concentric	CONC	Hexagon	HEX	Photograph	PHOTO	Tooth	T
Cotter	COT	Hydraulic	HYD	Point	PT	Total	TOT
Counterbore	CBORE	Illustrate	ILLUS	Point of		Typical	TYP
Countersink	CSK	Install	INSTL	intersection	PI	Vertical	VERT
		Intersect	INT	Point of		Washer	WASH
		Key	K	tangency	PT	Width	W

APPENDIX D

Scales and Grids

The method for developing scales and grids was discussed in Chapter Three. Over a period of time you will no doubt develop scales and grids of your own. For your use, two sets of axonometric scales and grids and one perspective grid have been included in this appendix.

The scales can be cut out, mounted on appropriately stiff material, and assembled. They will then function as a traditional triangular scale. The block diagram represents both the axis scales and inclination, and the correct ellipse is noted for each face. The axonometric triangles can be copied, thermofaxed as a transparency,

and mounted on thick acetate. As such they can be used as both a drawing and measuring device.

Many illustrators like to have the correct grid under the paper even though they use scales. This aids in both speed and in visualizing the three dimensions of the drawing.

D–1. Isometric grid.

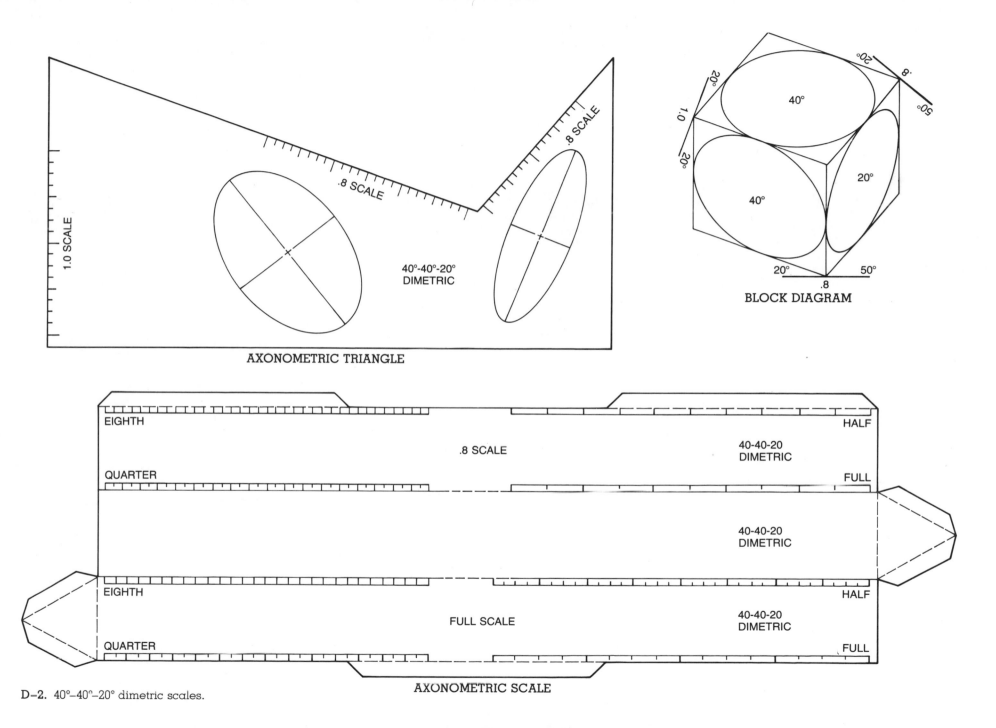

1.0 SCALE

.8 SCALE

.8 SCALE

40°-40°-20°
DIMETRIC

AXONOMETRIC TRIANGLE

.8

20°

1.0

20°

40°

20°

50°

.8

40°

20°

40°

20°

50°

.8

BLOCK DIAGRAM

EIGHTH

.8 SCALE

40-40-20
DIMETRIC

HALF

QUARTER

FULL

40-40-20
DIMETRIC

EIGHTH

HALF

FULL SCALE

40-40-20
DIMETRIC

QUARTER

FULL

AXONOMETRIC SCALE

D–2. 40°–40°–20° dimetric scales.

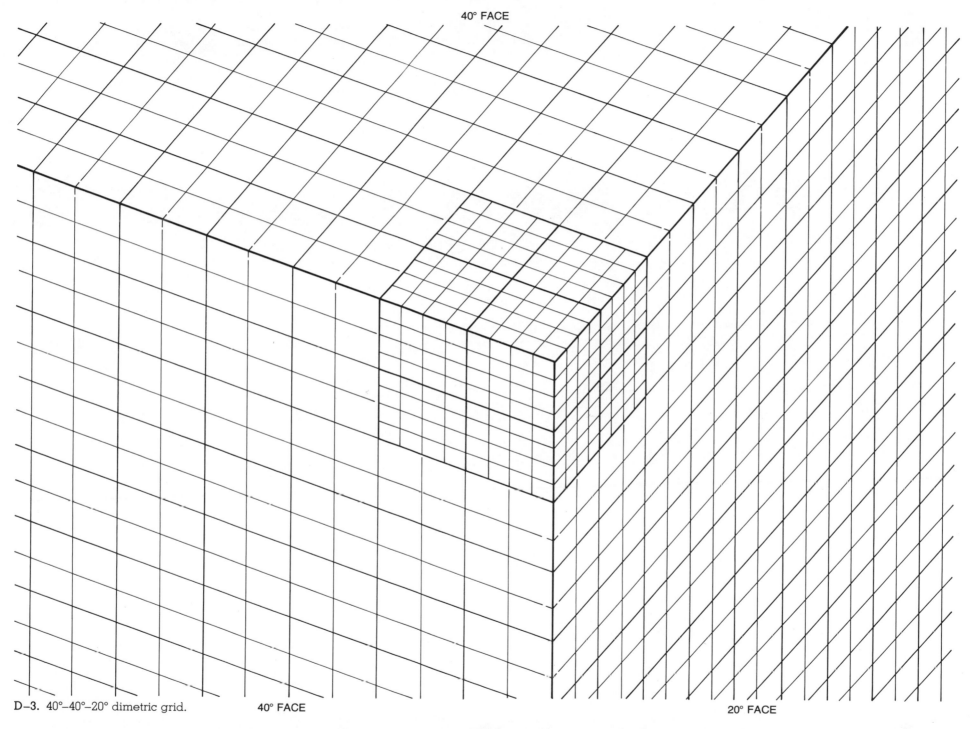

40° FACE

40° FACE

20° FACE

D–3. 40°–40°–20° dimetric grid.

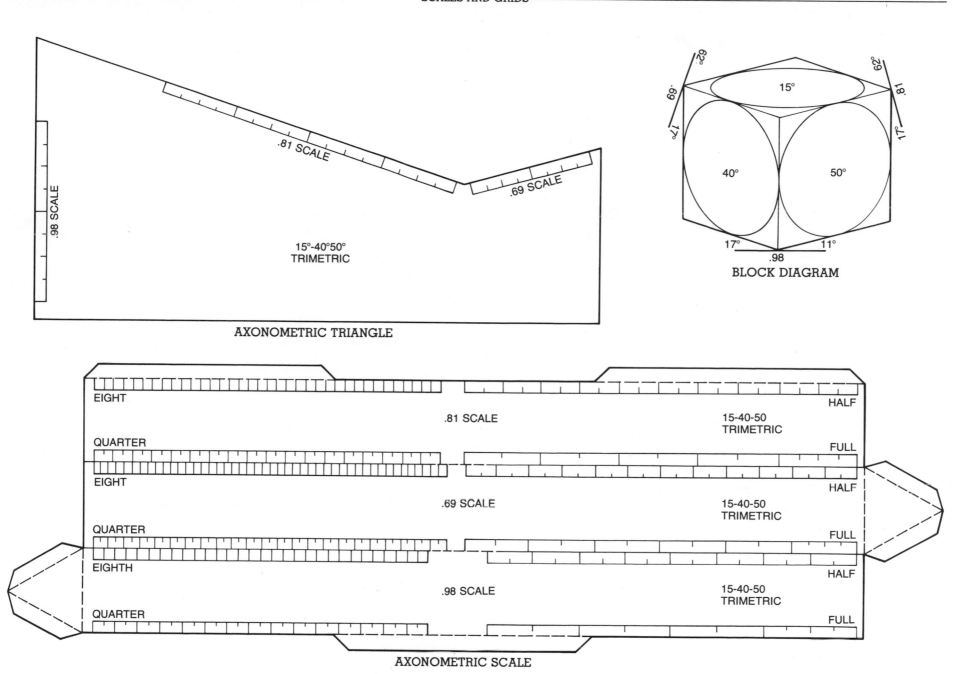

AXONOMETRIC TRIANGLE

BLOCK DIAGRAM

AXONOMETRIC SCALE

D–4. 15° 40°–50° trimetric scales.

15° FACE

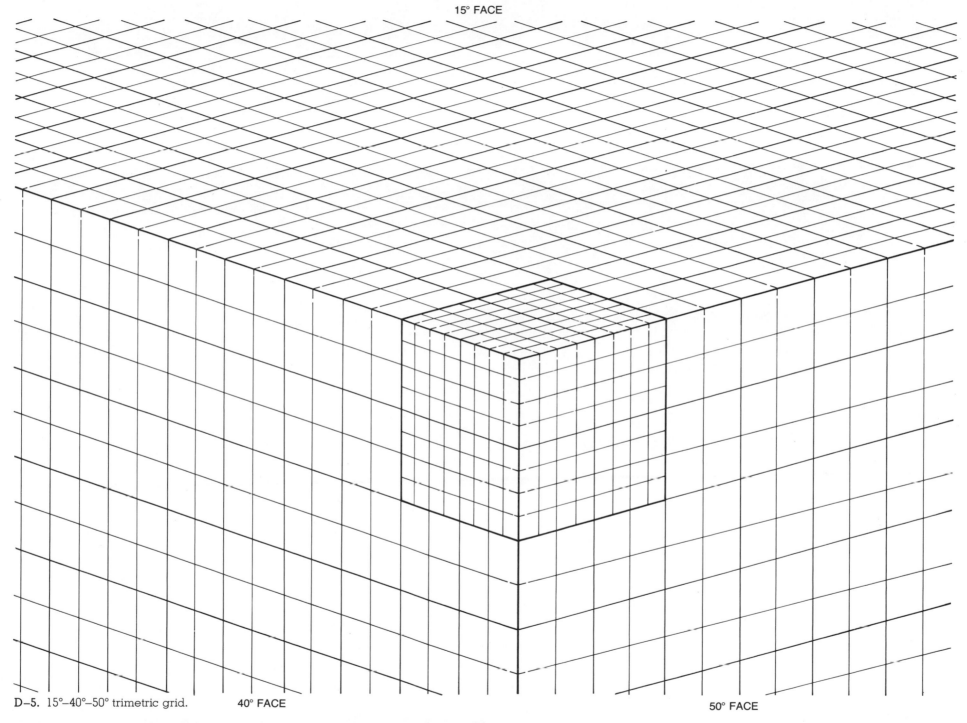

D–5. 15°–40°–50° trimetric grid.

40° FACE

50° FACE

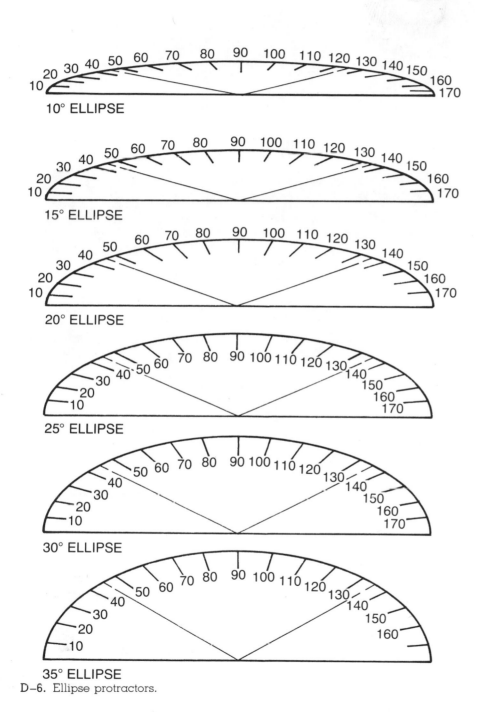

10° ELLIPSE

15° ELLIPSE

20° ELLIPSE

25° ELLIPSE

30° ELLIPSE

35° ELLIPSE

40° ELLIPSE

45° ELLIPSE

50° ELLIPSE

60° ELLIPSE

D–6. Ellipse protractors.

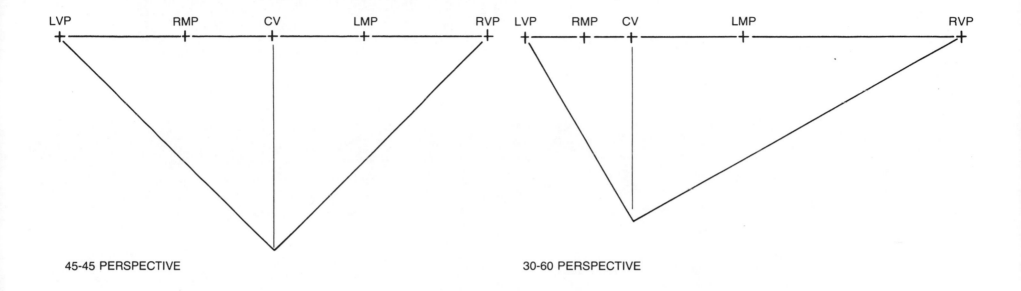

45-45 PERSPECTIVE

30-60 PERSPECTIVE

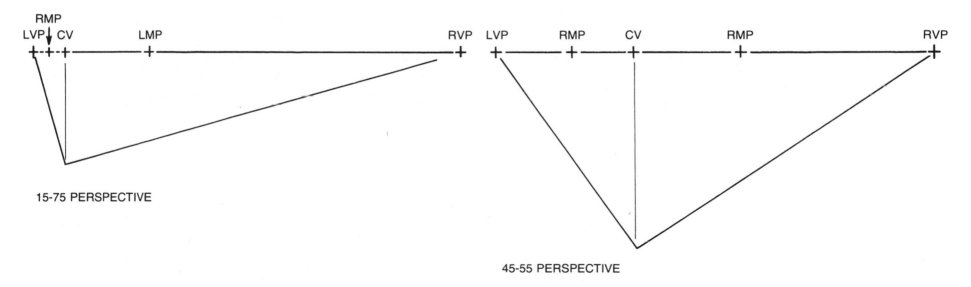

15-75 PERSPECTIVE

45-55 PERSPECTIVE

D–7. Perspective measuring point diagrams.

References

Some of the following references are not readily available or may be out of print. They are, however, commonly found in offices and schools.

VISUALIZATION

Rapid Viz. Hanks, Kurt, and Belliston, Larry. Los Altos, CA: William Kaufman, Inc., 1980.

Experiences in Visual Thinking. McKim, Robert. Monterey, CA: Brooks/Cole Publishing Company, 1980.

ILLUSTRATION

Technical Illustration. Thomas, T.A. New York: McGraw-Hill, 1978.

Scale Drawing. Nicyper, Raymond. Westport, CN: Graphicraft, 1973.

Illustrator's Ellipse Tips. Blakeslee, H.W. Baltimore, OH: Timely Productions, 1968.

Technical Illustration. Earle, James H. College Station, TX: Creative Publishing, 1978.

Estimating Illustration Costs—A Guide. Washington, DC: Society for Technical Communications, 1973.

Technical Illustrating. Gibby, Joseph C. Chicago, IL: American Technical Society, 1962.

Industrial Production Illustration. Hoelscher, R.P., Springer, C.H., and Pohle, R.F. New York: McGraw-Hill, 1946.

Technical Illustration. Morris, George E. Englewood Cliffs, NJ: Prentice-Hall, 1975.

ENGINEERING GRAPHICS

Engineering Drawing & Graphic Technology. French, T.E., and Vierck, C.J. New York: McGraw-Hill, 1978.

Graphics in Engineering Design. Levens, A., and Chalk, W. New York: John Wiley & Sons, 1980.

Engineering Drawing and Design. Jensen, C., and Hesel, J. New York: McGraw-Hill, 1979.

True Position Drafting Handbook. Troy, MI: Graphic Standard Institute, 1977.

Index